# EXPLORING THE UNIVERSE

**Anthea Maton**
Former NSTA National Coordinator
Project Scope, Sequence, Coordination
Washington, DC

**Jean Hopkins**
Science Instructor and Department Chairperson
John H. Wood Middle School
San Antonio, Texas

**Susan Johnson**
Professor of Biology
Ball State University
Muncie, Indiana

**David LaHart**
Senior Instructor
Florida Solar Energy Center
Cape Canaveral, Florida

**Maryanna Quon Warner**
Science Instructor
Del Dios Middle School
Escondido, California

**Jill D. Wright**
Professor of Science Education
Director of International Field Programs
University of Pittsburgh
Pittsburgh, Pennsylvania

Prentice Hall
Englewood Cliffs, New Jersey
Needham, Massachusetts

*Prentice Hall Science*

# Exploring the Universe

**Student Text and Annotated Teacher's Edition**
**Laboratory Manual**
**Teacher's Resource Package**
**Teacher's Desk Reference**
**Computer Test Bank**
**Teaching Transparencies**
**Product Testing Activities**
**Computer Courseware**
**Video and Interactive Video**

The illustration on the cover, rendered by Keith Kasnot, depicts Pluto's large moon, Charon, looming over Pluto's icy horizon.

Credits begin on page 167.

SECOND EDITION

© 1994, 1993 by Prentice-Hall, Inc., Englewood Cliffs, New Jersey 07632.

ISBN 0-13-401134-1

9 10    97 96 95

Prentice Hall
A Division of Simon & Schuster
Englewood Cliffs, New Jersey 07632

## STAFF CREDITS

| | |
|---|---|
| **Editorial:** | Harry Bakalian, Pamela E. Hirschfeld, Maureen Grassi, Robert P. Letendre, Elisa Mui Eiger, Lorraine Smith-Phelan, Christine A. Caputo |
| **Design:** | AnnMarie Roselli, Carmela Pereira, Susan Walrath, Leslie Osher, Art Soares |
| **Production:** | Suse F. Bell, Joan McCulley, Elizabeth Torjussen, Christina Burghard |
| **Photo Research:** | Libby Forsyth, Emily Rose, Martha Conway |
| **Publishing Technology:** | Andrew Grey Bommarito, Deborah Jones, Monduane Harris, Michael Colucci, Gregory Myers, Cleasta Wilburn |
| **Marketing:** | Andrew Socha, Victoria Willows |
| **Pre-Press Production:** | Laura Sanderson, Kathryn Dix, Denise Herckenrath |
| **Manufacturing:** | Rhett Conklin, Gertrude Szyferblatt |

### Consultants

| | |
|---|---|
| Kathy French | National Science Consultant |
| Jeannie Dennard | National Science Consultant |
| Brenda Underwood | National Science Consultant |
| Janelle Conarton | National Science Consultant |

## Contributing Writers

**Linda Densman**
*Science Instructor*
*Hurst, TX*

**Linda Grant**
*Former Science Instructor*
*Weatherford, TX*

**Heather Hirschfeld**
*Science Writer*
*Durham, NC*

**Marcia Mungenast**
*Science Writer*
*Upper Montclair, NJ*

**Michael Ross**
*Science Writer*
*New York City, NY*

## Content Reviewers

**Dan Anthony**
*Science Mentor*
*Rialto, CA*

**John Barrow**
*Science Instructor*
*Pomona, CA*

**Leslie Bettencourt**
*Science Instructor*
*Harrisville, RI*

**Carol Bishop**
*Science Instructor*
*Palm Desert, CA*

**Dan Bohan**
*Science Instructor*
*Palm Desert, CA*

**Steve M. Carlson**
*Science Instructor*
*Milwaukie, OR*

**Larry Flammer**
*Science Instructor*
*San Jose, CA*

**Steve Ferguson**
*Science Instructor*
*Lee's Summit, MO*

**Robin Lee Harris Freedman**
*Science Instructor*
*Fort Bragg, CA*

**Edith H. Gladden**
*Former Science Instructor*
*Philadelphia, PA*

**Vernita Marie Graves**
*Science Instructor*
*Tenafly, NJ*

**Jack Grube**
*Science Instructor*
*San Jose, CA*

**Emiel Hamberlin**
*Science Instructor*
*Chicago, IL*

**Dwight Kertzman**
*Science Instructor*
*Tulsa, OK*

**Judy Kirschbaum**
*Science/Computer Instructor*
*Tenafly, NJ*

**Kenneth L. Krause**
*Science Instructor*
*Milwaukie, OR*

**Ernest W. Kuehl, Jr.**
*Science Instructor*
*Bayside, NY*

**Mary Grace Lopez**
*Science Instructor*
*Corpus Christi, TX*

**Warren Maggard**
*Science Instructor*
*PeWee Valley, KY*

**Della M. McCaughan**
*Science Instructor*
*Biloxi, MS*

**Stanley J. Mulak**
*Former Science Instructor*
*Jensen Beach, FL*

**Richard Myers**
*Science Instructor*
*Portland, OR*

**Carol Nathanson**
*Science Mentor*
*Riverside, CA*

**Sylvia Neivert**
*Former Science Instructor*
*San Diego, CA*

**Jarvis VNC Pahl**
*Science Instructor*
*Rialto, CA*

**Arlene Sackman**
*Science Instructor*
*Tulare, CA*

**Christine Schumacher**
*Science Instructor*
*Pikesville, MD*

**Suzanne Steinke**
*Science Instructor*
*Towson, MD*

**Len Svinth**
*Science Instructor/*
*Chairperson*
*Petaluma, CA*

**Elaine M. Tadros**
*Science Instructor*
*Palm Desert, CA*

**Joyce K. Walsh**
*Science Instructor*
*Midlothian, VA*

**Steve Weinberg**
*Science Instructor*
*West Hartford, CT*

**Charlene West, PhD**
*Director of Curriculum*
*Rialto, CA*

**John Westwater**
*Science Instructor*
*Medford, MA*

**Glenna Wilkoff**
*Science Instructor*
*Chesterfield, OH*

**Edee Norman Wiziecki**
*Science Instructor*
*Urbana, IL*

## Teacher Advisory Panel

**Beverly Brown**
*Science Instructor*
*Livonia, MI*

**James Burg**
*Science Instructor*
*Cincinnati, OH*

**Karen M. Cannon**
*Science Instructor*
*San Diego, CA*

**John Eby**
*Science Instructor*
*Richmond, CA*

**Elsie M. Jones**
*Science Instructor*
*Marietta, GA*

**Michael Pierre McKereghan**
*Science Instructor*
*Denver, CO*

**Donald C. Pace, Sr.**
*Science Instructor*
*Reisterstown, MD*

**Carlos Francisco Sainz**
*Science Instructor*
*National City, CA*

**William Reed**
*Science Instructor*
*Indianapolis, IN*

## Multicultural Consultant

**Steven J. Rakow**
*Associate Professor*
*University of Houston—*
*Clear Lake*
*Houston, TX*

## English as a Second Language (ESL) Consultants

**Jaime Morales**
*Bilingual Coordinator*
*Huntington Park, CA*

**Pat Hollis Smith**
*Former ESL Instructor*
*Beaumont, TX*

## Reading Consultant

**Larry Swinburne**
*Director*
*Swinburne Readability*
*Laboratory*

# CONTENTS

## EXPLORING THE UNIVERSE

**CHAPTER 1**

**Stars and Galaxies** .................................10

1–1 A Trip Through the Universe.....................**12**
1–2 Formation of the Universe ......................**22**
1–3 Characteristics of Stars ..........................**28**
1–4 A Special Star: Our Sun ........................**38**
1–5 The Evolution of Stars ...........................**42**

**CHAPTER 2**

**The Solar System** .................................**54**

2–1 The Solar System Evolves ......................**56**
2–2 Motions of the Planets ...........................**60**
2–3 A Trip Through the Solar System .............**64**
2–4 Exploring the Solar System .....................**91**

**CHAPTER 3**

**Earth and Its Moon**...........................**102**

3–1 The Earth in Space...............................**104**
3–2 The Earth's Moon.................................**113**
3–3 The Earth, the Moon, and the Sun .........**120**
3–4 The Space Age ....................................**126**

*SCIENCE GAZETTE*

*Ian K. Shelton Discovers an Exploding Star*....**136**
*Looking for Life Beyond Earth: Should the
   Search for ETs Go On?*................................**138**
*Voyage to the Red Planet: Establishing the
   First Colony on Mars*.....................................**140**

# Activity Bank/Reference Section

| | |
|---|---|
| **For Further Reading** | 144 |
| **Activity Bank** | 145 |
| **Appendix A:** The Metric System | 152 |
| **Appendix B:** Laboratory Safety: Rules and Symbols | 153 |
| **Appendix C:** Science Safety Rules | 154 |
| **Appendix D:** Map Symbols | 156 |
| **Appendix E:** Star Charts | 157 |
| **Glossary** | 161 |
| **Index** | 164 |

# Features

**Laboratory Investigations**

| | |
|---|---|
| Identifying Substances Using a Flame Test | 50 |
| Constructing a Balloon Rocket | 98 |
| Observing the Apparent Motion of the Sun | 132 |

**Activity: Discovering**

| | |
|---|---|
| Using Star Charts | 12 |
| An Expanding Universe | 26 |
| Absolutely Apparent | 33 |
| Ellipses | 60 |
| Demonstrating Inertia | 62 |
| Orbital Velocities | 79 |
| Temperature, Daylight, and the Seasons | 106 |
| Lines of Force | 111 |

**Activity: Doing**

| | |
|---|---|
| Designer Constellations | 14 |
| Tube Constellations | 16 |
| Investigating Inertia | 63 |
| Build a Greenhouse | 70 |
| Planetary Sizes | 85 |
| Solar System I.Q. | 88 |
| A Foucault Pendulum Model | 104 |
| Observing the Moon | 120 |

**Activity: Calculating**

| | |
|---|---|
| Fusion Power | 37 |
| Comparing Diameters | 76 |
| The Earth on the Move | 108 |

**Activity: Thinking**

| | |
|---|---|
| Red or Blue Shift? | 24 |
| Outer Planetary Weather | 83 |

**Activity: Writing**

| | |
|---|---|
| A Wondrous Journey | 29 |
| Word Search | 40 |
| A New Planetary System? | 57 |

| | |
|---|---|
| An Important Theory | 59 |
| A Different Viewpoint | 96 |
| Space Technology Facilities | 131 |

**Activity: Reading**

| | |
|---|---|
| Psychohistory | 17 |
| Astrohistory | 90 |
| A Voyage to the Moon | 114 |

**Problem Solving**

| | |
|---|---|
| A Question of Intensity | 38 |
| What Causes Summer? | 107 |

**Connections**

| | |
|---|---|
| Signs of the Zodiac | 21 |
| The First Magellan Probe | 97 |
| Solar Wind Blows Out the Lights | 113 |

**Careers**

| | |
|---|---|
| Astronomer | 30 |
| Aerospace Engineer | 119 |

# CONCEPT MAPPING

Throughout your study of science, you will learn a variety of terms, facts, figures, and concepts. Each new topic you encounter will provide its own collection of words and ideas—which, at times, you may think seem endless. But each of the ideas within a particular topic is related in some way to the others. No concept in science is isolated. Thus it will help you to understand the topic if you see the whole picture; that is, the interconnectedness of all the individual terms and ideas. This is a much more effective and satisfying way of learning than memorizing separate facts.

Actually, this should be a rather familiar process for you. Although you may not think about it in this way, you analyze many of the elements in your daily life by looking for relationships or connections. For example, when you look at a collection of flowers, you may divide them into groups: roses, carnations, and daisies. You may then associate colors with these flowers: red, pink, and white. The general topic is flowers. The subtopic is types of flowers. And the colors are specific terms that describe flowers. A topic makes more sense and is more easily understood if you understand how it is broken down into individual ideas and how these ideas are related to one another and to the entire topic.

It is often helpful to organize information visually so that you can see how it all fits together. One technique for describing related ideas is called a **concept map**. In a concept map, an idea is represented by a word or phrase enclosed in a box. There are several ideas in any concept map. A connection between two ideas is made with a line. A word or two that describes the connection is written on or near the line. The general topic is located at the top of the map. That topic is then broken down into subtopics, or more specific ideas, by branching lines. The most specific topics are located at the bottom of the map.

To construct a concept map, first identify the important ideas or key terms in the chapter or section. Do not try to include too much information. Use your judgment as to what is really important. Write the general topic at the top of your map. Let's use an example to help illustrate this process. Suppose you decide that the key terms in a section you are reading are School, Living Things, Language Arts, Subtraction, Grammar, Mathematics, Experiments, Papers, Science, Addition, Novels. The general topic is School. Write and enclose this word in a box at the top of your map.

SCHOOL

Now choose the subtopics—Language Arts, Science, Mathematics. Figure out how they are related to the topic. Add these words to your map. Continue this procedure until you have included all the important ideas and terms. Then use lines to make the appropriate connections between ideas and terms. Don't forget to write a word or two on or near the connecting line to describe the nature of the connection.

Do not be concerned if you have to redraw your map (perhaps several times!) before you show all the important connections clearly. If, for example, you write papers for Science as well as for Language Arts, you may want to place these two subjects next to each other so that the lines do not overlap.

One more thing you should know about concept mapping: Concepts can be correctly mapped in many different ways. In fact, it is unlikely that any two people will draw identical concept maps for a complex topic. Thus there is no one correct concept map for any topic! Even

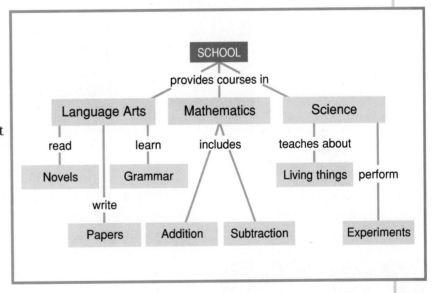

though your concept map may not match those of your classmates, it will be correct as long as it shows the most important concepts and the clear relationships among them. Your concept map will also be correct if it has meaning to you and if it helps you understand the material you are reading. A concept map should be so clear that if some of the terms are erased, the missing terms could easily be filled in by following the logic of the concept map.

# EXPLORING THE UNIVERSE

The Trifid Nebula is a huge mass of glowing gas and stars. It contains many young stars that generate a great deal of heat and light. The red portions of the nebula are mainly glowing hydrogen gas. The blue regions contain mostly dust particles that reflect light from the stars in the nebula. ▶

A nearly full moon was photographed by the Apollo 13 astronauts. ▼

Set aglow by fiery stars deep within its core, the gases of the Trifid Nebula sweep out into the blackness of space. The wispy red and blue cloud balloons out to a size that is almost unimaginable. It could gobble up thousands and thousands of solar systems! But it is so far away from the Earth that people can enjoy its beauty without fearing it will engulf them.

Although this dazzling cloud of gas is about 5000 light-years, or 50,000 trillion kilometers, from the Earth, it is one of the Earth's neighbors in space. The mysterious object astronomers call the Trifid Nebula is part of the Earth's starry neighborhood, the Milky Way Galaxy. Equally mysterious, but closer to home, are the objects with which the Earth and its moon share the solar system: the sun, the

# CHAPTERS

**1** Stars and Galaxies     **3** Earth and
**2** The Solar System        Its Moon

planets and their moons, meteors, asteroids, and
comets. In the pages that follow, you will explore
these and other objects in space and perhaps
begin to unravel some of their mysteries.

*The rings of Saturn and
six of its many moons
can be seen in this
composite photograph.*

# Discovery *Activity*

## *Colors of the Rainbow*

Is white light really white? Find out for yourself.

1. Hold a prism in front of a sheet of white paper.

2. Shine a beam of white light through the prism onto the paper.
   What happens to the white light as it passes through the
   prism? How is this similar to a rainbow?

   ■ What do you think causes the colors of a rainbow?

   ■ What do you think would happen if you were to pass the
   beam of light through a second prism? Try it.

# Stars and Galaxies

## Guide for Reading

*After you read the following sections, you will be able to*

**1–1 A Trip Through the Universe**
- Describe the groups into which stars are classified.
- Describe the Milky Way galaxy.

**1–2 Formation of the Universe**
- Describe the big-bang theory and relate it to the formation of the universe.
- Explain how red shift is used to determine the movements of galaxies.

**1–3 Characteristics of Stars**
- Classify stars by size, mass, color, temperature, and brightness.

**1–4 A Special Star: Our Sun**
- Describe the four main layers of the sun.

**1–5 The Evolution of Stars**
- Describe the life cycles of stars.

Where does the sun fit into the universe? In 1914, American astronomer Harlow Shapley asked himself this question. Until then most astronomers believed that the sun was the center of the entire universe. Shapley, however, was not so sure.

Shapley began studying large groups, or clusters, of stars. He made a model showing where these clusters were located. Using his model, Shapley found that the clusters were grouped together in a gigantic sphere whose center was thousands of light-years from the sun. Shapley believed the center of this gigantic sphere was the real center of the universe. If so, our sun and nearby stars were actually near the edge of the universe!

Most of the stars that were known in Shapley's time were in a part of the sky called the Milky Way. The Greek name for the Milky Way is *galaxias kylos*, which means milky circle. So Shapley's universe came to be called the galaxy. Shapley believed that all the matter in the universe was located in this single galaxy. Outside the galaxy there was empty space.

But was Shapley right? Had he found the center of the universe? To shed some light on the answer, read on.

## Journal *Activity*

***You and Your World*** What would it be like to experience the wonders of outer space? In your journal, write a letter to NASA describing why you should be selected for the first trip to a distant star.

*Notice the bright stars shining amidst the glowing gas and dust of the Orion nebula.*

## Guide for Reading

*Focus on these questions as you read.*

▶ What are galaxies? What are the three main types of galaxies?

▶ What is the size and shape of the Milky Way galaxy?

## ACTIVITY

### DISCOVERING

*Using Star Charts*

**1.** Turn to the star chart appendices at the back of this textbook.

**2.** Take the star chart outside on a starry night. Make sure you hold the star chart in the proper position for the time and date you are observing.

**3.** Make a drawing of what you see.

**4.** Repeat this activity in about one month.

How many constellations can you find in your first drawing? In your second drawing?

■ Are there any other changes in your drawing? If so, how can you explain them?

# 1–1 A Trip Through the Universe

Look up at the stars on a clear moonless night. Hundreds, perhaps thousands, of stars fill the sky. Each of these twinkling lights is actually a sun—a huge sphere of hot, glowing gas. Many of these suns are much larger and brighter than our own sun. But even a person with extremely good eyesight can see only a tiny fraction of the stars in the entire universe. Telescopes, of course, reveal many more stars. Most of the stars are so far away, however, that they cannot be seen individually, even with the most powerful telescope. Fortunately, although astronomers are unable to locate most individual stars, they can detect huge groups of stars.

To better understand what astronomers have learned about the stars, you can begin with a quick journey through the known universe. Most scientists believe that nothing can travel faster than the speed of light, which is about 300,000 kilometers per second. That's fast, but not fast enough for your trip. For even at the speed of light a journey through the universe would take billions of years. So in the spaceship of your mind you will have to travel much faster, stopping only occasionally to view some of the wonders of space.

**Figure 1–1** *Only a small portion of the billions upon billions of stars in the universe are shown in this photograph.*

## Multiple-Star Systems

You begin your trip by heading directly toward the star closest to the sun, Alpha Centauri. Although it is "close" by space standards, Alpha Centauri is still about 4.3 light-years from Earth.

Because our sun is a single-star system, astronomers believed for many years that most stars form individual star systems. For example, Alpha Centauri, when viewed from Earth, appears as a single speck in the sky. As you approach Alpha Centauri, however, you quickly discover that early astronomers were wrong. Alpha Centauri is not a single star at all, but three stars that make up a triple-star system! So Alpha Centauri is a multiple-star system. In fact, only one of the stars, called Proximi Centauri, is actually the closest to Earth.

As you continue your journey, you begin to realize that Alpha Centauri is not unusual. You discover that about half the stars in the sky have at least one companion star. Most of these stars are double-star systems in which two stars revolve around each other. Double-star systems are called **binary stars** (the prefix *bi-* means two). See Figure 1–2.

Thousands of years ago Arab shepherds discovered that about every three days a certain bright star suddenly became dim and disappeared, only to brighten again. In fear of this strange star, they

**Figure 1–2** *This photograph shows Sirius, the Dog Star, and its small binary companion star. What are binary stars?*

**Figure 1–3** *Algol is a binary star system. Each time the large dark star passes between the bright star and Earth, the bright star appears to dim and disappear. When does it reappear?*

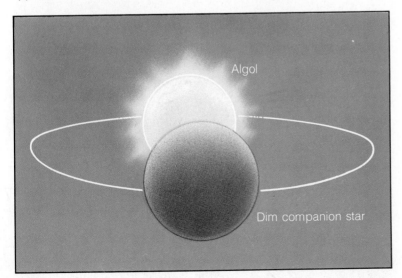

Algol

Dim companion star

## ACTIVITY DOING

*Designer Constellations*

Go outside on a starry night and draw the placement of the stars you can see in your area.

Use your drawing to design a new constellation. Your constellation can be based on a mythological creature or something from modern day.

named it Algol, the "ghoul." Since your journey steers you by Algol, you will be able to discover the reason for Algol's winking on and off. Algol is a binary-star system. One of Algol's stars is a small, bright-blue star. It is visible from Earth. The other star is a large, dim, yellow star. It is not visible from Earth, so the Arabs could not have known of its existence. About every three days the large star passes between the smaller star and Earth, blocking off some of the smaller star's light. So every three days the smaller star appears to disappear. Can you explain why it reappears again?

## Constellations: Star Groups That Form Patterns

From Algol you can continue your journey in any direction. One path may take you past the Dog Star—Sirius—which is more than 8 light-years from Earth. Another path may take you to the North Star—Polaris—more than 700 light-years from Earth. Polaris has long been an important star to navigators at sea because they knew if they steered toward Polaris they were heading north.

**Figure 1–4** *These are some of the constellations you can see in the night sky. What do the pointer stars point to?*

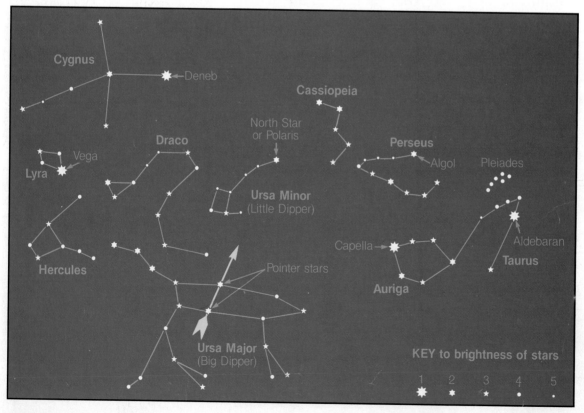

Polaris is at the end of the handle of a group of stars called the Little Dipper. The Little Dipper, in turn, makes up the **constellation** of stars called Ursa Minor, or the Little Bear. Constellations are groups of stars in which people at one time thought they saw imaginary figures of animals or people. See Figure 1–4.

One of the best-known constellations is the Big Bear, or Ursa Major. The seven stars in the back end and tail of the Big Bear form the Big Dipper, which can be easily seen in the northern sky. Two bright stars in the cup of the Big Dipper are known as the pointer stars because they point toward Polaris.

On clear winter nights, you can see the large constellation Orion, the Hunter. There are two bright stars in this constellation: Betelgeuse (BEET-uhl-jooz) and Rigel (RIGH-juhl). Nearby are other constellations: Gemini, Canis Major, or the Big Dog, and Canis Minor, or the Little Dog. Some of the summer constellations that are easy to recognize are Scorpius, Leo, and Virgo. What constellations do you know?

## Novas

No time to dally with the constellations, as there is much more to see. The Little Dipper fades into the distance as you steer your ship toward a web of glowing stars. Then, without any warning, a tiny star seems to explode in a burst of light. You are fortunate indeed. For you have witnessed a rare event in space—you have seen a **nova.**

A nova is a star that suddenly increases in brightness up to 100 times in just a few hours or days. Soon after it brightens, the nova slowly begins to grow dim again. Astronomers believe that almost all novas are members of binary-star systems. Gases from the companion star in the system occasionally strike the surface of the nova star. When this happens, a nuclear explosion results, and heat, light, and gases burst into space. See Figure 1–5.

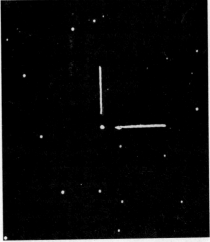

**Figure 1–5** *In 1935, a star in the constellation Hercules erupted in a rare nova. A month later the same star, shown by the arrows, had returned to normal.*

Figure 1–6 *The Pleiades is an open cluster of stars (right). The Hercules cluster of stars is globular (left).*

## ACTIVITY

### DOING

*Tube Constellations*

**1.** Obtain a thick cardboard tube, such as a mailing tube, a flashlight, a straight pin, tape, and black construction paper.

**2.** With the pin, punch out a constellation on the black construction paper. Use Figure 1–4 as a guide. *Hint:* You may want to use different size pins to show stars of different brightness.

**3.** Tape the construction paper over one end of the tube.

**4.** Insert a flashlight into the opposite end of the tube and project the constellation onto a flat, dark surface. Challenge your classmates to identify your tube constellations.

## Star Clusters

When the light from the nova finally fades, you are able again to pay close attention to the mass of stars you are approaching. When you do, you notice that although there are stars almost everywhere you can see, a great many stars seem to be grouped in huge clusters.

There are two types of star clusters. See Figure 1–6. Open clusters such as the Pleiades are not well organized and contain hundreds of stars. Globular clusters, which are more common, are arranged in a spherical, or round, shape. Globular clusters, such as the cluster in the constellation Hercules, contain more than 100,000 stars.

Star clusters appear to the unaided eye on Earth as one star or as a faint, white cloud. Why do you think that you cannot see the individual stars in a cluster?

## Nebulae

Once you have passed what seems like a thousand or more globular clusters, you begin to notice that the number of stars is thinning. You are rapidly approaching an area of space that seems empty. But before you get there, it is time to put on a special pair of glasses. Up until now you have seen with your eyes alone, so you have been limited to seeing only the visible light rays your eyes can detect. The special glasses will allow you to see all the different kinds of rays stars can give off.

**Figure 1–7** *Most stars are born in the gas and dust that make up a nebula. The photographs show the Red Nebula (right) and the Tarantula Nebula (left).*

You are completely unprepared for the spectacle that awaits you when you put on your glasses. To your left you notice a mass of brilliant stars that shine mainly with X-rays. Ahead you see stars beaming ultraviolet light at you. Strangely enough, almost all the stars seem to be giving off some radio waves as well. You wonder briefly why you have never heard the stars on your radio. Then you remember that astronomers do just that when they "listen" to the sky with radio telescopes.

Of all the views your glasses give you, there is one that seems most exciting. Now that you can see infrared rays—or heat rays—an entirely new universe is revealed. And nothing could be more spectacular than the huge clouds of dust and gas you see glowing between the stars. Each massive cloud, probably the birthplace of new stars, is called a **nebula.** See Figure 1–7.

## Galaxies

You began this chapter by learning how astronomer Harlow Shapley believed the entire universe could be found in one huge **galaxy,** the Milky Way. In 1755, long before Shapley was born, the German philosopher-scientist Immanuel Kant suggested a different theory. Kant believed that the sun was part of a vast galaxy, but that there were other "island universes," or galaxies, scattered throughout space as well.

Your ship has finally reached the area of space that seemed empty before. You are about to leave the Milky Way galaxy. Was Shapley right—will there

A**CTIVITY**

**READING**

*Psychohistory*

Interested in societies of the future when people have moved to distant planets? Want to know how science and history can be combined into a single subject that can be used to rule the universe? If so, then you are guaranteed to enjoy the *Foundation Series* by Isaac Asimov.

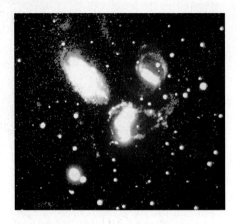

**Figure 1–8** *Each galaxy in this cluster of galaxies might hold several trillion or more stars. Astronomers believe there may be over 100 billion galaxies in the universe.*

**Figure 1–9** *Three different spiral galaxies are shown in these photographs. What is the major feature of a spiral galaxy?*

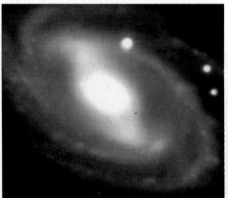

be nothing but empty space from here on? Or was Kant correct and will there be countless more worlds to visit? Seconds later the answer is clear. Before you, stretching out in every direction, thousands and thousands of galaxies shine in an awesome display of size and power. Even though the Milky Way is so large that it would take light more than 100,000 years to travel from one end to another, it is but one tiny galaxy in a sea of galaxies. **Galaxies, which contain various star groups, are the major features of the universe.** In fact, astronomers now believe there may be more than 100 billion major galaxies, each with billions of stars of its own! See Figure 1–8.

You cannot visit more than a handful of galaxies on your journey, but it really doesn't matter. For you quickly find that most galaxies fit one of three types. Many galaxies are **spiral galaxies.** See Figure 1–9. Spiral galaxies, such as Andromeda, which is 2 million light-years from Earth, are shaped like pinwheels. They have huge spiral arms that seem to reach out into space, ready to grab passing visitors that stray too close. The Milky Way, in which planet Earth is located, is another example of a spiral galaxy.

Galaxies that vary in shape from nearly spherical to flat disks are called **elliptical galaxies.** These galaxies contain very little dust and gas. The stars in elliptical galaxies are generally older than those in

other types of galaxies. This should not be surprising, since you learned that stars are born in huge clouds of gas and dust (nebula), which are rare in elliptical galaxies.

The third type of galaxy does not have the orderly shape of either the elliptical or spiral galaxies. These galaxies are called irregular galaxies. Irregular galaxies have no definite shape. The Large and Small Magellanic Clouds are irregular galaxies. They are the closest galaxies to the Milky Way. Several hundred irregular galaxies have been observed by astronomers, but they are much less common than spiral or elliptical galaxies.

## The Milky Way Galaxy

It is almost time for our journey to end. You have traveled in a huge circle. In the distance you see the familiar spiral shape of your own galaxy, the Milky Way. In your mind, at least, you are the first person to observe the Milky Way from outside the galaxy. From this distance you can see that the Milky Way is a huge pinwheel-shaped disk with a bulge in the center. See Figure 1–11.

Most of the older stars in the Milky Way are found near the nucleus, or center, of the galaxy. The stars there are crowded together thousands of times more densely than in the spiral arms. This nucleus,

**Figure 1–10** *Elliptical galaxies, such as the giant M87 galaxy, are oval-shaped and do not contain spiral arms (top). Irregular galaxies have no definite shape (bottom).*

**Figure 1–11** *From the side, the Milky Way appears to be a narrow disk with a bulge in the center. Seen from the front, the galaxy reveals its spiral structure.*

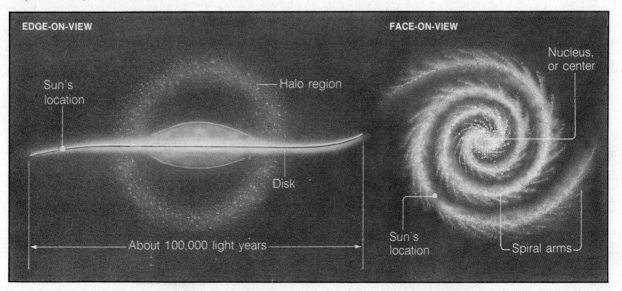

EDGE-ON-VIEW

FACE-ON-VIEW

Nucleus, or center

Sun's location

Halo region

Disk

Sun's location

Spiral arms

About 100,000 light years

almost 20,000 light-years across, is hidden from our view on Earth by thick clouds of hot dust and gases.

Scientists estimate the Milky Way to be about 100,000 light-years in diameter and about 15,000 light-years thick. So even at the speed of light, it would take 100,000 years to travel across the Milky Way! If you could spot the light from Earth's sun, shining among the 100 billion stars in the Milky Way, you would notice that the sun is located in one of the pinwheel's spiral arms, almost 30,000 light-years from the central bulge. Earth's sun, along with many of the stars in the spiral arms, is among the younger stars in the Milky Way.

As you approach the Milky Way from above, you notice that all its many stars are rotating counter-clockwise about its center. In fact, scientists estimate that it takes our sun and its planets about 200 million years to rotate once about the center of the galaxy.

Well, your trip was an exciting one, but it is time to come back to Earth. You have traveled unimaginable distances in your mind and returned safely. Now it is time to look into how the extraordinarily beautiful and majestic universe formed. For the remainder of your study of stars and galaxies, you will have to travel through the pages of this textbook. But don't worry—it won't be any more dangerous than your trip through the universe.

**Figure 1–12** *This edge-on-view of the Milky Way was made by plotting the locations of over 7000 known stars. Where is the sun located in the Milky Way?*

# CONNECTIONS

## Signs of the Zodiac

About 2000 years ago, astronomers wondered what the sky would look like if the stars could be seen during the day. Based on their observations of the night sky, some astronomers determined that during the daytime the sun would appear to move across the sky, entering a different constellation each month. These twelve constellations, one per month, came to be called the *zodiac*. Each constellation was called a sign of the zodiac. Many ancient people believed that the month a person was born in, and hence that person's sign, influenced the person's behavior, emotions, and even his or her fate. Even today, thousands of years later, some people still believe in the powers of the zodiac and read their horoscope daily. Most people read their horoscope for fun, however, not because they believe it to be true.

The illustration shows the original symbols for the twelve signs of the zodiac and the constellation each sign relates to. Many of these symbols may be familiar to you, as they are commonly used in jewelry. Under what sign were you born? Do you know how your emotions are supposed to be affected by your sign?

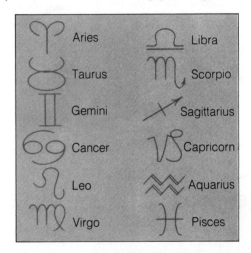

## 1-1 Section Review

1. Why are galaxies considered the major features of the universe?
2. Name and describe the three main types of galaxies.
3. Many binary stars are called eclipsing binaries. Explain why this term is appropriate. (*Hint:* Think about Algol when you answer this question.)

**Connections—*Science and Technology***
4. How has our ability to detect "invisible" forms of light contributed to our knowledge of the universe?

## Guide for Reading

*Focus on these questions as you read.*

▶ How does a spectroscope provide information about distant stars?

▶ According to the big-bang theory, how did the universe form?

# 1–2 Formation of the Universe

Astronomers use various telescopes to study stars. Optical telescopes detect visible light from stars. Radio telescopes detect radio waves emitted by stars. X-ray telescopes detect X-rays; ultraviolet telescopes detect ultraviolet rays. Finally, infrared telescopes examine infrared radiation from stars. Since stars give off some or all of these types of rays, telescopes are important tools for the astronomer.

Telescopes, however, are not the only tools of astronomers. An equally important tool is the **spectroscope.** Although visible light from stars appears white to your eyes, the light given off by stars usually contains a mixture of several different colors, all of which combine to make white light. A spectroscope can break up the light from a distant star into its characteristic colors. See Figure 1–14.

When light enters a spectroscope, the light is first focused into a beam by a lens. The beam of light then passes through a prism. A prism separates light into its different colors. (If you have a prism, you can prove this for yourself.) The band of colors formed when light passes through a prism is called

**Figure 1–13** *Telescopes have been developed to detect the many types of light stars emit—both visible and invisible. The photograph of the Andromeda Galaxy was taken using an infrared telescope (right). The image of the galaxy called Centaurus A was taken with a radio telescope (top left). Parts of the constellation Sagittarius are seen in this X-ray photograph (bottom left).*

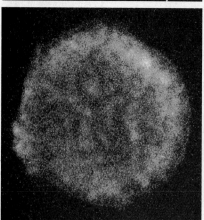

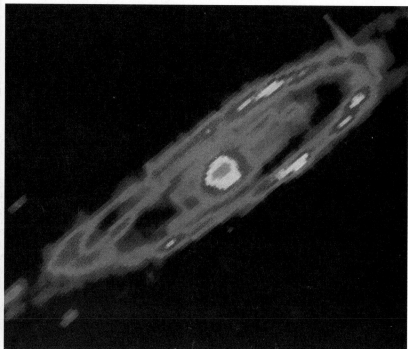

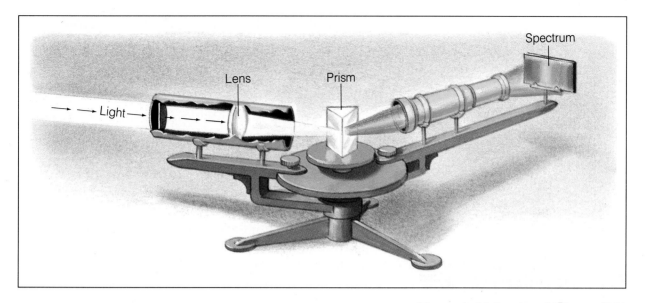

**Figure 1–14** *In a spectroscope, light passes through a prism and is broken into a band of colors called a spectrum.*

a **spectrum.** The kind of spectrum produced by the light from a star tells astronomers a lot about that star.

## Stars on the Move

Every single object in the universe is on the move. The moon, for example, moves around the Earth. The Earth, in turn, travels around the sun. The sun moves about the center of the Milky Way galaxy. As you have read, astronomers suggest there may be as many as 100 billion major galaxies. And like the other objects in space, each and every galaxy is on the move. By using a spectroscope, astronomers can determine whether a particular galaxy is moving toward the Earth or away from the Earth.

## The Red Shift

Drop a stone into a pool of water and you will see water waves traveling away from the stone in all directions. The distance from the top of one wave (crest) to the top of the next wave is called the wavelength.

Light from stars travels to Earth as light waves. You have read that a spectroscope breaks up light into a spectrum. This happens because each color of light has a different wavelength. When light strikes the prism in a spectroscope, the prism bends the

**A**ctivity Bank

All the Colors of the Rainbow, p.146

**Figure 1–15** *Despite the vastness of space, galaxies do collide on occasion, as demonstrated by this rare photograph of two galaxies colliding. Such collisions may last many millions of years.*

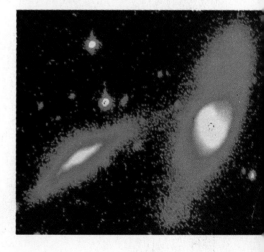

light according to the wavelength of each color. Some wavelengths are bent more than others by the prism. So the white light that enters the prism comes out as a band of colors. Each color has a different wavelength.

Suppose a star is rapidly approaching the Earth. The light waves from the star will be compressed, or pushed together. In fact, wavelengths from an approaching star often appear shorter than they really are. Shorter wavelengths of light are characteristic of blue and violet light. So the entire spectrum of an approaching star appears to be shifted slightly toward the blue end of the spectrum. This shifting is called the blue shift.

If a star is moving away from the Earth, the light waves will be slightly expanded as they approach the Earth. The wavelengths of the light will appear longer than they really are. Longer wavelengths of light are characteristic of the red end of the spectrum. So the spectrum of a star moving away from the Earth appears to be shifted slightly toward the red end. This is called the **red shift.** Astronomers know that the more the spectrum of light is shifted toward the blue or red end of the spectrum, the faster the star is moving toward or away from the Earth.

The apparent change in the wavelengths of light that occurs when an object is moving toward or away from the Earth is called the **Doppler effect.** You have probably "heard" another kind of Doppler effect right here on Earth. If you are in a car at a railroad crossing when a train is approaching, the first sound of the train's whistle will be high-pitched. The sound of the whistle will become low-pitched as the train passes by and moves away from you. In this example, the Doppler effect involves sound waves. But the same principle applies to light waves moving toward or away from Earth. See Figure 1–16.

When astronomers first used the spectroscope to study the light from stars in distant galaxies, they had a surprise. None of the light from distant galaxies showed a blue shift. That is, none of the galaxies was moving toward the Earth. Instead, the light from every distant galaxy showed a red shift. Every galaxy in the universe seemed to be moving away from the Earth.

# ACTIVITY

## THINKING

*Red or Blue Shift?*

The top spectrum represents a star not moving toward or away from Earth. The bottom two spectra show what would happen if the star were moving with respect to Earth.

**1.** Compare the bottom two spectra with the top spectrum.

**2.** Determine which spectrum is of a star moving toward Earth and which is of a star moving away.

Explain your answer in terms of red and blue shift.

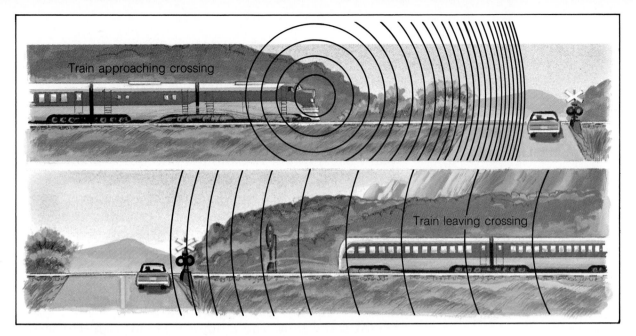

**Figure 1–16** *As the train approaches the crossing (top), sound waves are crowded together and reach the listener's ears with a high pitch. As the train leaves the crossing (bottom), sound waves are farther apart and have a lower pitch. What term is used to describe this effect?*

After examining the red shifts of distant galaxies, astronomers concluded that the universe is expanding. Galaxies near the edge of the universe are racing away from the center of the universe at tremendous speeds. Galaxies closer to the center are also moving outward, but at slower speeds. What can account for an expanding universe?

## The Big-Bang Theory

Astronomers believe that the expanding universe is the result of an enormous and powerful explosion called the big bang. The **big-bang theory** may explain how the universe formed. **The big-bang theory states that the universe began to expand with the explosion of concentrated matter and energy and has been expanding ever since.** According to the theory, all the matter and energy in the universe was once concentrated into a single place. This place, of course, was extremely hot and dense. Then some 15 to 20 billion years ago, an explosion—the big bang—shot the concentrated matter and energy in all directions. The fastest moving matter traveled farthest away. Energy, too, began moving away from the area of the big bang.

**Figure 1–17** *What does this illustration tell you about galaxies?*

If the big-bang theory is correct, the energy left from the big bang will be evenly spread out throughout the universe. This energy is known as background radiation. And indeed, scientists have discovered that the background radiation is almost the same throughout the entire universe. This constant background radiation is one observation that supports the big-bang theory.

After the initial big bang, the force of **gravity** began to affect the matter racing outward in every direction. Gravity is a force of attraction between objects. All objects have a gravitational attraction for other objects. The more massive the object is, the stronger its gravitational attraction. This force of gravity began to pull matter into clumps.

At some time, the clumps formed huge clusters of matter. These clumps became the galaxies of the universe. But even as the galaxies were forming, the matter inside the galaxies continued to race away from the area where the big bang had occurred. And this is just what astronomers have discovered. All of the galaxies are speeding away from the center of the universe.

## An Open Universe

Most astronomers feel that the big-bang theory leads to two possible futures for the universe. Perhaps the galaxies will continue racing outward. In this case, the universe will continue to expand. Such a universe is called an open, eternal universe. But eternal does not mean "forever" when it comes to the universe. In an open universe, the stars will eventually die off as the last of their energy is released. So the future of an open universe is one in which there will be nothing left. An open universe leads to total emptiness. But even if the universe is open, its end will not occur for many billions of years.

## A Closed Universe

Most astronomers do not feel that the universe is an open universe. Instead, they suspect that the gravitational attraction between the galaxies will one day cause their movement away from each other to slow down. The expansion of the universe will finally

## Activity Bank

Swing Your Partner, p.147

# ACTIVITY

### An Expanding Universe

Use a balloon and small circles cut from sticky labels to make a model of an expanding universe. First, decide how to put your model together. What will represent the galaxies? How will the balloon enable you to show how the universe is expanding?

■ Do the galaxies get any larger as the universe expands?

■ What relationship exists between the speed of the galaxies moving apart and their initial distances from one another?

**Figure 1–18** *A closed universe will eventually contract. Scientists picture a closed universe as similar to the surface of a ball (top). An open universe will expand until all of the stars die off. Scientists picture an open universe in the shape of a saddle (bottom).*

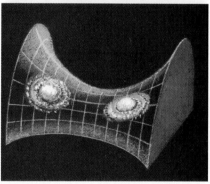

come to a halt. Then gravity will begin to pull the galaxies back toward the center of the universe. When this happens, every galaxy will begin to show a blue shift in its spectrum. Recall that a blue shift means that a galaxy is moving toward the Earth.

As the galaxies race back toward the center of the universe, the matter and energy will again come closer and closer to the central area. After many billions of years, all the matter and energy will once again be packed into a small area. This area may be no larger than the period at the end of this sentence. Then another big bang will occur. The formation of a universe will begin all over again. A universe that periodically expands and then contracts back on itself is called a closed universe. In a closed universe, a big bang may occur once every 80 to 100 billion years.

## Quasars

If the universe is expanding, then objects near the edge of the universe are the oldest objects in the universe. Put another way, these objects took longer to reach their current position than objects closer to the center of the universe did. The most distant known objects in the universe are about 12 billion light-years from Earth. They are called **quasars** (KWAY-zahrz). The word quasar stands for quasi-stellar radio sources. The prefix *quasi-* means something like. The word *stellar* means star. So a quasar is a starlike object that gives off radio waves.

Quasars are among the most studied, and the most mysterious, objects in the universe. They give off mainly radio waves and X-rays. The mystery of quasars is the tremendous amount of energy they give off. Although they may seem too small to be galaxies, they give off more energy than 100 or more galaxies combined!

If the big-bang theory is correct, quasars at the edge of the universe were among the first objects formed after the big bang. In fact, scientists now believe that quasars may represent the earliest stages in

**Figure 1–19** *This quasar, seen through an X-ray telescope, is some 12 billion light-years from Earth. How long does it take the quasar's light to reach Earth?*

the formation of a galaxy. So when scientists observe quasars, they are observing the very edge and the very beginning of the universe. Keep in mind that the light from a quasar 12 billion light-years from the Earth has traveled more than 12 billion years to reach the Earth. Astronomers observing quasars are, in a sense, looking back into time.

## 1-2 Section Review

1. How does a spectroscope enable astronomers to determine the characteristics of distant stars and galaxies?
2. Briefly describe the big bang. What two pieces of information provide evidence of the big bang?
3. What role did gravity play in the formation of galaxies?

**Connection—*Social Studies***

4. The German philosopher Nietzsche went mad because he believed that he had already taken every action he took. How did the concept of a closed universe affect Nietzche's beliefs?

**Guide for Reading**

*Focus on this question as you read.*

▶ *How do the size, mass, color, temperature, and brightness of stars vary?*

# 1-3 Characteristics of Stars

Astronomers estimate that there may be more than 200 billion billion stars in the universe. **Stars differ in many features, including size, mass, color, temperature, and brightness.** It might seem an impossible task for astronomers to study so many different stars. While stars do vary in a great many ways, however, there are certain forces that govern all stars. By studying the stars that can be examined in detail and the forces that they must obey, astronomers gain knowledge about the vast numbers of stars they cannot closely observe.

## How Large Are Stars?

Most stars are so far away that they appear as tiny points of light through even the most powerful

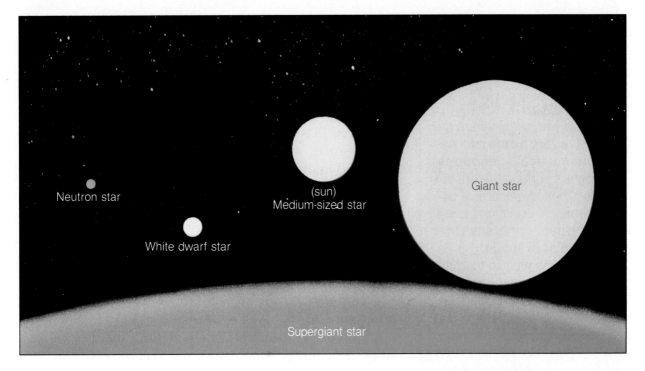

telescopes. However, looks can be deceiving. For stars actually vary tremendously in size. Astronomers have divided stars into five main groups by size. See Figure 1–20.

Our sun has a diameter of about 1,392,000 kilometers, or about 109 times the diameter of Earth. That may seem enormous to you, but the sun is actually a medium-sized star. Medium-sized stars make up the majority of the stars you can see in the sky. They vary in size from about one-tenth the size of the sun to about ten times its size. Many of these stars are very bright. Sirius, for example, is about twice the diameter of the sun and is the brightest star in the night sky.

Stars with diameters about 10 to 100 times as large as the sun are called **giant stars**. For example, the diameter of the orange giant Aldebaran is 45 times the sun's diameter. Even the giant stars, however, seem tiny in comparison to the largest of all stars—the **supergiant stars**. Supergiants have diameters up to 1000 times the diameter of the sun. Some of the best-known supergiants are Rigel, Betelgeuse, and Antares. To get some idea of just how large these stars can be, suppose the red supergiant Antares were to replace the sun. It would burn Earth to cinders; in fact, it would extend well

**Figure 1–20** *Stars come in a variety of sizes. What are the largest stars called? The smallest?*

*A Wondrous Journey*

Write a story describing the wondrous sights you would see if you could take a journey from one end of the Milky Way to the other. In your story use at least five words from the vocabulary list at the end of the chapter.

**Figure 1–21** *Scientists cannot take a bit of a star from the Large Magellanic Cloud in order to study the star's composition. How do astronomers know what elements make up the stars in the Magellanic Cloud?*

beyond planet Mars. Supergiants, however, pay a price for their huge size. They die off quickly and are the shortest-lived stars in the universe.

Not all stars are larger than the sun. Some, such as **white dwarfs,** are even smaller than Earth. The smallest known white dwarf, Van Maanen's star, has a diameter that is less than the distance across the continent of Asia.

The smallest stars of all are called **neutron stars**. A typical neutron star has a diameter of only about 16 kilometers. That may be less than the total distance you travel to and from school.

## Composition of Stars

Suppose you were given an unknown substance and asked to discover what it was made of. You might start by looking closely at the object from all sides. You might touch it to see how it feels and smell it to see if you could recognize its aroma. You would probably want to perform some other tests on it as well.

Astronomers cannot take a bit of a star and test it to see what it is made of. But they can determine the composition of stars, even stars many light-years from Earth. To determine the composition of a star, astronomers turn to the spectroscope.

How can a spectroscope show what a star is made of? Let's begin with the simple example of ordinary table salt. Table salt, or sodium chloride, is made of the elements sodium and chlorine. If table salt is placed in a flame, the flame will burn bright yellow.

The yellow flame is caused by the sodium in salt. A yellow flame, then, is a characteristic of the element sodium.

Suppose the yellow light from the flame is passed through a spectroscope. No matter how many times this is done, two thin lines will always appear in the spectrum produced by the spectroscope. These two thin lines will always appear in exactly the same place in the spectrum. See Figure 1–22. In fact, no other element will produce the same two lines as sodium. In a way, these two lines are the "fingerprint" of the element sodium.

Other elements also produce a characteristic set of lines when they are heated and the light given off is passed through a spectroscope. So every known element has a fingerprint. By passing the light from a star through a spectroscope, astronomers can determine exactly what elements are in that star. How? They compare the spectral lines from the star to the spectral lines, or fingerprints, of the known elements.

By using the spectroscope, astronomers have found that almost all stars have the same general composition. The most common element in stars is hydrogen. Hydrogen is the lightest element. It makes up 60 to 80 percent of the total mass of a star. Helium is the second most common element in a typical star. It is the second lightest element. In fact, the combination of hydrogen and helium makes up about 96 to 99 percent of a star's mass. All other elements in a star total little more than 4 percent of the star's mass. These other elements often include oxygen, neon, carbon, and nitrogen.

**Figure 1–22** *With the spectrum produced by a spectroscope, scientists can identify the elements in distant stars. These spectra are of the elements sodium (top), hydrogen (center), and helium (bottom).*

## Surface Temperature of Stars

Are you familiar with the heating coil on top of an electric stove? When the stove is off, the coil is dark. Then, when the stove is turned on, the coil begins to change color. Soon the coil is bright red, and you know that it is very hot. As you can see, the color a hot object gives off is a good indicator of its temperature.

The sun is a yellow star. But stars come in many other colors. By studying the color of a star, astronomers can determine its surface temperature.

| STAR COLORS AND SURFACE TEMPERATURES | | |
|---|---|---|
| Color | Average Surface Temperatures (°C) | Examples |
| Blue or blue-white | 35,000 | Zeta Eridani Spica Algol |
| White | 10,000 | Sirius Vega |
| Yellow | 6000 | Procyon Sun Alpha Centauri A |
| Red-orange | 5000 | Alpha Centauri B |
| Red | 3000 | Proxima Centauri Barnard's star |

**Figure 1–23** *How do scientists use color to determine the surface temperature of distant stars?*

Keep in mind that the surface temperature of a star is much lower than the temperature in the star's center, or core. For example, the sun has a surface temperature of about 6000°C. Yet the temperature of the sun's core can reach 15,000,000°C.

Using color as their guide, astronomers have determined that the surface temperature of the hottest stars is about 50,000°C. Such stars shine with a blue-white light. Red stars, which are among the coolest stars, have a surface temperature of about 3000°C. Most other stars have surface temperatures between these two extremes. See Figure 1–23.

## Brightness of Stars

Suppose you were given a small flashlight and a large spotlight. You know, of course, that the spotlight is much bigger and brighter than the flashlight is. However, if a friend held the dim flashlight about a meter away from you and another friend held the spotlight a long distance away, the flashlight would appear brighter. So the brightness you saw would depend on the strength of the light, the size of the light source, and the distance the light source was from your eyes.

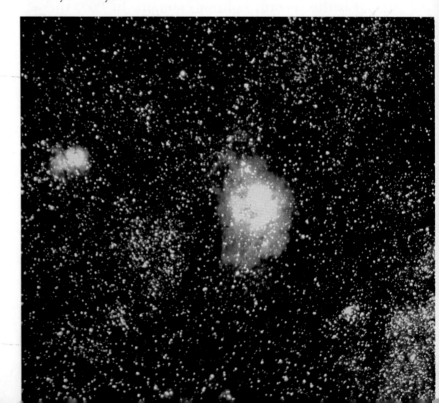

**Figure 1–24** *The brightness of these stars in the Trifid Nebula, as seen from Earth, is called absolute brightness. What term is used for the actual brightness of a particular star?*

The brightness of a star depends on its size, its surface temperature, and its distance from Earth. When you look at the night sky, some stars appear brighter than others. As in the case of the flashlight and the spotlight, however, the star that appears brighter may not really be brighter at all. So astronomers call the brightness of a star as it appears from Earth its **apparent magnitude.**

If astronomers could take two stars and place them exactly the same distance from Earth, they could easily tell which star was really brighter. Astronomers cannot move stars, of course, but they can calculate a star's actual brightness. The amount of light a star actually gives off is called its **absolute magnitude.**

The brightness of most stars is constant. Some stars, however, vary in brightness and are called variable stars. One type of variable star changes size as well as brightness in regular cycles. Stars of this type are called pulsating variable stars. The North Star, Polaris, for example, changes from bright to dim and back again in a four-day cycle. Astronomers call these pulsating stars Cepheid (SEF-ee-id) variables because the first one was discovered in a group of stars called Cepheus.

## The Hertzsprung-Russell Diagram

In the early 1900s, Danish astronomer Ejnar Hertzsprung and American astronomer Henry Norris Russell, working independently, found a relationship between the absolute magnitude and the temperature of stars. They discovered that as the absolute magnitude of stars increases, the temperature usually also increases. The relationship between the absolute magnitude and the surface temperature is shown in Figure 1–25 on page 34. You can see a definite pattern. This pattern forms the **Hertzsprung-Russell (H-R) diagram**. The Hertzsprung-Russell diagram is the single most important diagram astronomers use today.

On the Hertzsprung-Russell diagram, the surface temperature of stars is plotted along the horizontal axis. The absolute magnitude, or actual brightness, of stars is plotted along the vertical axis. If you study Figure 1–25, you will see that most stars fall in an

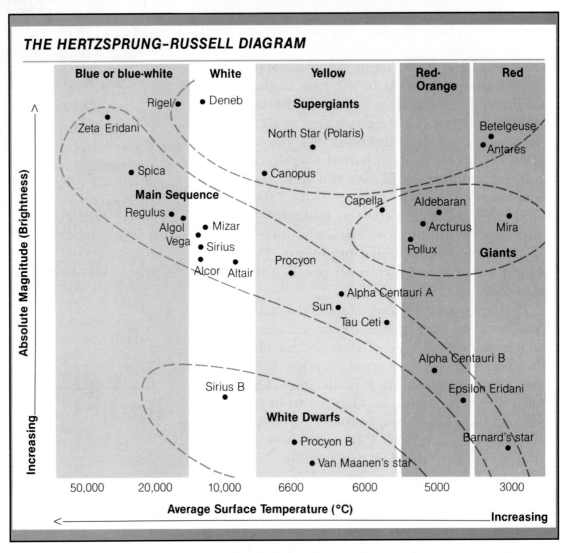

## THE HERTZSPRUNG-RUSSELL DIAGRAM

**Figure 1–25** *The Hertzsprung-Russell diagram shows that, for most stars, as the absolute magnitude increases, the surface temperature also increases.*

area from the upper left corner to the lower right corner. This area is called the main sequence. The stars within this area are called **main-sequence stars.**

Main-sequence stars make up more than 90 percent of the stars in the sky. The hottest main-sequence stars shine with a blue or blue-white light and are located in the upper left corner of the H-R diagram. Cool, dim main-sequence stars appear in the lower right corner.

The Hertzsprung-Russell diagram also identifies the other 10 percent of stars. These stars are no longer on the main sequence. They have changed to their present condition as they have aged. In the area above the main sequence are stars called red giants and supergiants. In the area below the main sequence are the white dwarfs. These white dwarfs are smaller and dimmer than main-sequence stars.

## Measuring Star Distance

Since we cannot travel to stars and there certainly aren't any tape measures long enough to stretch anyway, how do astronomers determine the distance to different stars? Actually, it's a pretty complicated problem and sometimes some guesswork is involved.

One method of measuring the distance to stars is called **parallax** (PAR-uh-laks). Parallax refers to the apparent change in the position of a star in the sky. This apparent change in position is not due to the movement of the star. Instead, it is due to the change in the Earth's position as the Earth moves around the sun. So the star stays in the same place and the Earth moves.

In Figure 1–26, you can see how parallax is used to determine the distance to a star. First, the apparent position of the star in June and in December is noted. A line is then drawn between the Earth's position in these months and the center of the sun. This straight line will become the base of a triangle. The length of this base line is known to astronomers because it has already been carefully measured.

Next, a diagonal line is drawn from each end of the base line to the apparent position of the star in June and in December. These three lines form a triangle. The tip of the triangle is the true position of the star. Then a vertical line is drawn from the true position of the star to the base of the triangle. This line, labeled X, is the actual distance to the star. Since astronomers can determine the angles within the parallax triangle, they can calculate the length of line X. In this way, they can measure the true distance from the Earth to the star.

Parallax is a reliable method for measuring the distance to stars relatively close to Earth. The distance to stars more than 100 light-years away, however, cannot be found using parallax. Why? The angles within the parallax triangle are too small to be accurately measured.

To determine the distance to a star more than 100 light-years away, astronomers use the brightness

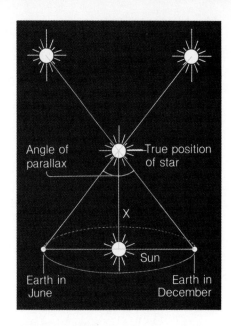

**Figure 1–26** *Scientists can use an apparent change in position called parallax to measure star distance. By calculating the length of the line marked X, they can find the actual distance to the star.*

of the star. They plug the star's apparent magnitude and its absolute magnitude into a complicated formula. The formula provides a close approximation of the distance to that star.

Neither brightness nor parallax will work when a star is more than 7 million light-years from Earth—and most stars are at least that far away. To determine the distance to these stars, astronomers once again use the spectroscope. As you have read, light from a star moving away from Earth has a red shift in its spectrum. Astronomers measure the amount of red shift in a star's spectrum and use complex formulas to calculate how far from Earth a star is located. This method of calculating star distances is controversial, and not all scientists agree as to just how far away many stars are from Earth.

## Why Stars Shine

You have learned how light from a star can be used to determine its composition, surface temperature, and distance. But what exactly causes a star to shine? To answer this question, you must look deep into the core of the star.

Within the core of a star, gravitational forces are extremely strong. In fact, gravity pulls together the

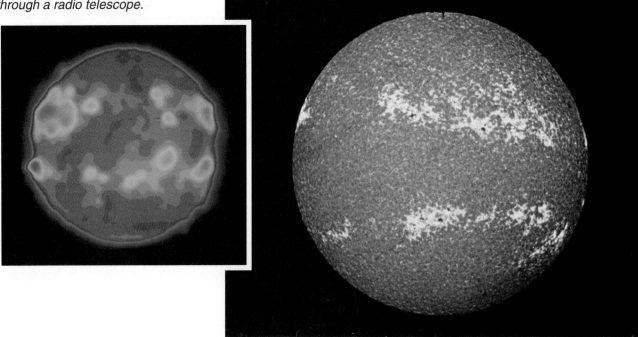

**Figure 1–27** *Although we think of the sun as giving off only visible light, these photographs prove us wrong. The photograph on the right is a combination of an X-ray and ultraviolet light. The photograph on the left shows our sun as seen through a radio telescope.*

atoms of hydrogen gas in the core so tightly that they become fused together. This process is called **nuclear fusion.** During nuclear fusion, hydrogen atoms are fused to form helium atoms. Nuclear fusion, then, allows a star to produce a new element by combining other elements.

The sun changes about 600 billion kilograms of hydrogen into 595.8 billion kilograms of helium every second. As you can see from these numbers, during this fusion process 4.2 billion kilograms of the original mass of the hydrogen seem to be lost every second. Magic? Not really. The missing mass has been changed into energy. Most of the rest of the mass is changed into heat and light. And that is why a star shines. Of course, not all the light from nuclear fusion is visible light. Some of it may be infrared radiation, ultraviolet radiation, radio waves, and X-rays.

Back on Earth, nuclear fusion can be both constructive and destructive. The most destructive force known is the hydrogen bomb, in which hydrogen atoms are fused to form helium atoms, and huge amounts of energy are released. But one day nuclear fusion may become the most constructive force known. For controlled nuclear fusion in a nuclear-power plant would provide unlimited energy that is relatively pollution-free. Scientists hope that sometime in the next century nuclear-fusion power plants may solve much of Earth's energy needs.

*Fusion Power*

Our sun changes 600 billion kilograms of hydrogen into 595.8 billion kilograms of helium every second. The remaining 4.2 billion kilograms are changed into the energy that pours out from the sun.

Determine how many grams of hydrogen are converted into energy in one minute. In one hour.

## 1–3 Section Review

1. Describe how stars vary in size, composition, temperature, color, mass, and brightness.
2. How do astronomers determine the surface temperature of stars?
3. Compare absolute magnitude and apparent magnitude.

**Critical Thinking—*Making Inferences***
4. Why does the parallax method of measuring star distances require observations of a star made six months apart?

# PROBLEM Solving

## A Question of Intensity

Light intensity can be simply defined as the amount of light that falls on a given area. The intensity of a star is of great interest to astronomers. But astronomers are not the only ones who need to know about light intensity. Anyone who uses a camera understands the importance of light intensity on the quality of his or her photographs and how it affects the type of film that must be used. But how does the intensity of light change as you move farther away from a light source?

### Relating Cause and Effect

Using a light source, a light meter, and a one-meter cardboard square, determine how light intensity changes over distance. Explain the results of your experiment (*Hint:* The total amount of light does not change over distance.)

---

## Guide for Reading

*Focus on this question as you read.*

▶ *What are the four main layers of the sun, and how do they differ?*

# 1–4 A Special Star: Our Sun

About 150 million kilometers from Earth there is a very important star—our sun. The sun is not unusual compared to other stars in the universe. It is a medium-sized, middle-aged yellow star about 4.6 billion years old. But without the sun, there would be no life on Earth.

## Layers of the Sun

The sun is a ball-shaped object made of extremely hot gases. It is an average star in terms of size, temperature, and mass. It measures 1.35 million

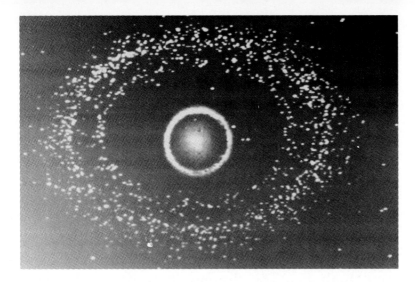

**Figure 1–28** *This computer-enhanced photograph surprised astronomers when it revealed a giant ring of dust circling the sun. The ring is almost 1.5 million kilometers from the sun.*

kilometers in diameter. If the sun were hollow, more than 1 million planet Earths could fit inside it! Although the sun's volume is more than 1 million times greater than that of Earth, its density is only one quarter that of Earth. Why do you think the Earth is more dense than the sun?

Since the sun is made only of gases, there are no clear boundaries within it. But four main layers can be distinguished. **Three layers make up the sun's atmosphere, and one layer makes up its interior.** See Figure 1–29.

**Figure 1–29** *The three main layers of the sun's atmosphere are the corona, the chromosphere, and the photosphere. What is the hottest part of the sun?*

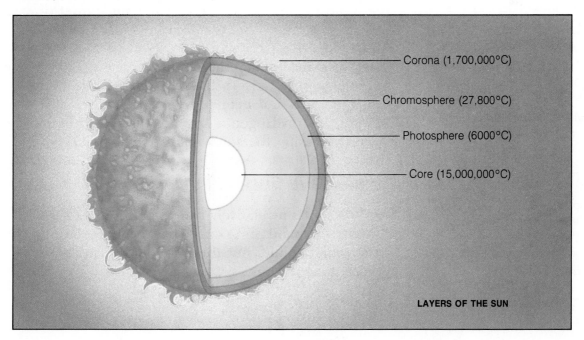

Corona (1,700,000°C)

Chromosphere (27,800°C)

Photosphere (6000°C)

Core (15,000,000°C)

**LAYERS OF THE SUN**

**Figure 1–30** *The sun's corona becomes visible during a total solar eclipse. What object in space blocks out the rest of the sun during a solar eclipse?*

**A**ctivity Bank

How Can You Observe the Sun Safely?, p.148

## **A**CTIVITY

### WRITING

*Word Search*

Have you ever wondered about the origins of words? Very often, especially in science, you can figure out the meaning of a word if you understand its parts. The terms *photosphere* and *chromosphere* are good examples.

Using reference materials, look up the meanings of the parts *chromo-, photo-,* and *-sphere.* How do they relate to the terms scientists use for parts of the sun's atmosphere?

**CORONA** The outermost layer of the sun's atmosphere is called the **corona** (kuh-ROH-nuh). Gas particles in the corona can reach temperatures up to 1,700,000°C. But if a spacecraft could pass through the corona and be shielded from the rest of the sun's heat, the temperature of the spacecraft would barely rise! The reason for this is simple. The gas particles in the corona are spread so far apart that not enough particles would strike the spacecraft at any one time to cause a rise in temperature.

**CHROMOSPHERE** Beneath the corona is the middle layer of the sun's atmosphere, the **chromosphere** (KROH-muh-sfir). The chromosphere is several thousand kilometers thick. But sometimes gases in the chromosphere suddenly flare up and stream as far as 16,000 kilometers into space. Temperatures in the chromosphere average 27,800°C.

**PHOTOSPHERE** The innermost layer of the sun's atmosphere is called the **photosphere.** The photosphere is about 550 kilometers thick and is often referred to as the surface of the sun. Temperatures in the photosphere usually do not exceed 6000°C.

**CORE** You may have noticed that the temperature decreases greatly from the corona through the photosphere. But the temperature begins to rise again in the interior of the sun. The interior of the sun includes all of the sun except the three layers of the atmosphere. At the edge of the sun's interior, near the photosphere, temperatures may reach 1 million degrees Celsius. But in the center of the sun, called the **core,** temperatures may reach up to 15 million degrees Celsius. It is here in the sun's core that hydrogen and helium gases churn in constant motion, and hydrogen atoms are fused into helium atoms, releasing the sun's energy as heat and light.

## The Active Sun

The sun is a relatively calm star compared to stars that expand and contract or erupt violently from time to time. But there is still a lot of activity going on in the sun.

**PROMINENCES** Many kinds of violent storms occur on the sun. One such solar storm is called

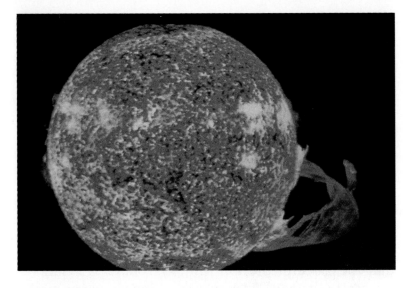

**Figure 1–31** *A huge solar prominence rises out of the sun like a twisted sheet of gas. What is a solar prominence?*

a **prominence** (PRAHM-uh-nuhns). Prominences are seen from Earth as huge bright arches or loops of gas. These twisted loops of hot gas usually originate in the chromosphere. Prominences sometimes bend backward and shower the gases back onto the sun. Other prominences from the chromosphere erupt to heights of a million kilometers or more above the sun's surface. During a solar prominence, gases and energy are sent into space. One incredibly large prominence, photographed on June 4, 1946, grew to almost the size of the sun in one hour before it disappeared a few hours later. Figure 1–31 shows a solar prominence shooting into space.

**SOLAR FLARES** Another kind of storm on the sun shows up as bright bursts of light on the sun's surface. These bursts of light are called **solar flares.** A solar flare usually does not last more than an hour. But during that time, the temperature of the solar-flare region can be twice that of the rest of the sun's surface. Huge amounts of energy are released into space during a solar flare.

**SOLAR WIND** A continuous stream of high-energy particles is released into space in all directions from the sun's corona. This stream is the **solar wind.** Solar flares sometimes increase the speed and strength of the solar wind. This increase in the solar wind can interfere with radio signals and telephone communications on Earth.

**SUNSPOTS** When astronomers observe the sun, they sometimes see dark areas on the sun's surface.

**Figure 1–32** *The green dots on the surface of the sun are colorized images of sunspots. What are sunspots?*

These dark areas are called **sunspots.** Sunspots appear dark because they are cooler than the rest of the sun's surface.

Sunspots are storms in the lower atmosphere of the sun. They may be as small as 16 kilometers in diameter or as large as 160,000 kilometers in diameter. The number of sunspots that appear on the sun at any one time is always changing. But periods of very active sunspot activity seem to occur every ten to eleven years. This activity interferes with communication systems on Earth.

Astronomers have observed that sunspots move across the sun's surface. This movement indicates that the gases in the sun spin, or rotate. The sun rotates on its **axis.** The axis is an imaginary vertical line through the center of the sun. Gases around the middle of the sun appear to rotate on the axis once every 25 days. But not all parts of the sun rotate at the same speed. Some parts of the sun take longer to rotate than others. What do you think accounts for this?

## 1–4 Section Review

1. List and describe the four main layers of the sun.
2. Describe four types of storms on the sun.
3. How have astronomers determined that the sun rotates on its axis?

**Connection—***Meteorology*
4. Explain the difference between the solar wind and winds on Earth.

**Guide for Reading**

*Focus on these questions as you read.*

▶ What is the life cycle of a star?

▶ How does the starting mass of a star relate to its evolution?

# 1–5 The Evolution of Stars

Does it surprise you that the title of this section is called The Evolution of Stars? If you are like most people, you may think of evolution as something that deals with changes in living things. The definition of evolution, however, can be thought of in simple terms as change over time. Using that definition, many things can be considered to evolve. Planet Earth, for example, has changed greatly since it

**Figure 1–33** *These rock formations in Arches National Park, Utah, demonstrate that planet Earth evolves, or changes, over time. What forces carved out the unusual rock formations?*

formed some 4.5 billion years ago. Rivers have carved canyons out of solid rock; plants have produced oxygen and turned a poisonous atmosphere into one in which animals can survive; mountains have risen, eroded away, and been carried as sediments to the sea.

Astronomers agree that stars also evolve, or change over time. The stars you see, including the sun, did not always look the way they do today. These stars will continue to change. Changes may take place over a few million years, or perhaps several billion years. Astronomers refer to the evolution of a star as the life cycle of a star.

Some stars have existed almost since the origin of the universe. Other stars, such as the sun, have come from the matter created by the first stars. From their studies of stars, astronomers have charted the life cycle of a star from its "birth" to its "death." According to the present theory of star evolution, the many different kinds of stars in the sky represent the various stages in the life cycle of a star.

## Protostars

You have read that galaxies contain huge clouds of dust and gases called nebulae. The most current theory of star formation states that new stars are born from the gases in a nebula. Over time, some of the hydrogen gas in a nebula is clumped together by gravity. The hydrogen atoms form a spinning cloud of gas within the nebula. Over millions of years more and more hydrogen gas is pulled into the spinning cloud. Collisions between hydrogen atoms become more frequent. These collisions cause the hydrogen gas to heat up.

**Figure 1–34** *New stars and protostars are forming today in the dust and gas that make up the Orion Nebula. The sun formed in a similar nebula over 5 billion years ago.*

When the temperature within the spinning cloud reaches about 15,000,000°C, nuclear fusion begins. The great heat given off during nuclear fusion causes a new star, or **protostar,** to form. As a result of nuclear fusion, the protostar soon begins to shine and give off heat and light. At that point, a star is born.

## Medium-Sized Stars

Once a protostar forms, its life cycle is fixed. Everything that will happen to that star has already been determined. **The main factor that shapes the evolution of a star is how much mass it began with.**

For the first few billion years, the new star continues to shine as its hydrogen is changed into helium by nuclear fusion in the star's core. But eventually most of the star's original supply of hydrogen is used up. By this time, most of the star's core has been changed from hydrogen to helium. Then the helium core begins to shrink. As it shrinks, the core heats up again. The outer shell of the star is still composed mainly of hydrogen. The energy released by the heating of the helium core causes the outer hydrogen shell to expand greatly. As the outer shell expands, it cools and its color reddens. At this point, the star is a red giant. It is red because cooler stars shine red. And it is a giant because the star's outer shell has expanded from its original size.

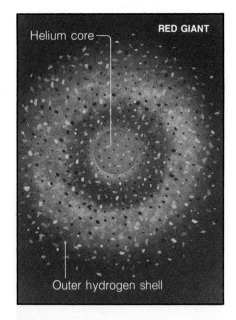

**Figure 1–35** *This illustration shows a red giant star. What does the red color of the outer shell indicate?*

RED GIANT

Helium core

Outer hydrogen shell

As the red giant ages, it continues to "burn" the hydrogen gas in its shell. The temperature within the helium core continues to get hotter and hotter too. At about 200,000,000°C, the helium atoms in the core fuse together to form carbon atoms. Around this time, the last of the hydrogen gas surrounding the red giant begins to drift away. This drifting gas forms a ring around the central core of the star. This ring is called a planetary nebula—although it has nothing to do with planets. See Figure 1–36.

At some point in the red giant's life, the last of the helium atoms in its core are fused into carbon atoms. The star begins to die. Without nuclear fusion taking place in its core, the star slowly cools and fades. Finally gravity causes the last of the star's matter to collapse inward. The matter is squeezed so tightly that the star becomes a tiny white dwarf.

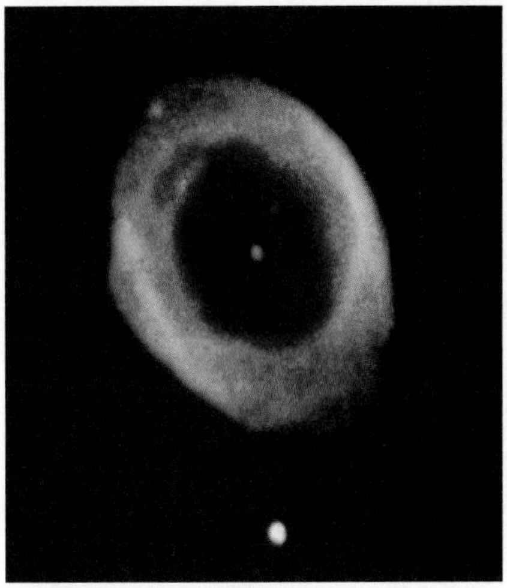

## White Dwarfs

The matter squeezed into a white dwarf is extremely dense. In fact, a single teaspoon of matter in a white dwarf may have a mass of several tons. But a white dwarf is not a dead star. It still shines with a hot white light.

At some point, the last of the white dwarf's energy is gone. It becomes a dead star. The length of time it takes a medium-sized star to become a white dwarf and die depends on the mass of the star when it first formed. It will take about 10 billion years for a medium-sized star such as the sun to evolve from formation to death. A smaller medium-sized star may take as long as 100 billion years. But a larger medium-sized star may die within only a few billion years. As you can see, the smaller the starting mass of a star, the longer it will live.

## Massive Stars

The evolution of a massive star is quite different from that of a medium-sized star. At formation, massive stars usually have at least six times as much mass as our sun. Massive stars start off like medium-sized stars. They continue on the same life-cycle path until they become red giants, or even supergiants. Unlike

**Figure 1–36** *This ring nebula, or planetary nebula, is all that is left of the gases that once surrounded a red giant star.*

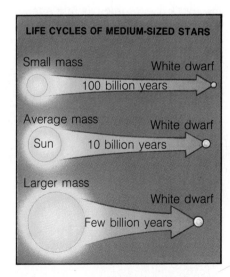

**Figure 1–37** *Medium-sized stars all end up as white dwarfs. What is the relationship between mass and the time it takes a medium-sized star to become a white dwarf?*

Figure 1–38 *These photographs show a star before and after a supernova. What characteristic of the star will determine its fate after the supernova?*

medium-sized stars, however, massive stars do not follow the path from red giant to white dwarf. They take a completely different path.

Recall that a red giant becomes a white dwarf when all the helium in its core has turned to carbon. In a massive star, gravity continues to pull together the carbon atoms in the core. When the core is squeezed so tightly that the heat given off reaches about 600,000,000°C, the carbon atoms begin to fuse together to form new and heavier elements such as oxygen and nitrogen. The star has begun to become a factory for the production of heavy elements. The core of the massive star is so hot that fusion continues until the heavy element iron forms. But not even the tremendous heat of the massive star can cause iron atoms to fuse together.

## Supernovas

By the time most of the nuclear fusion in a massive star stops, the central core is mainly iron. Although the process is not well understood, the iron atoms begin to absorb energy. Soon this energy is released, as the star breaks apart in a tremendous explosion called a **supernova.** A supernova can light the sky for weeks and appear as bright as a million suns. (Keep in mind that a supernova is very different from the nova you read about earlier. Only the names are similar.)

Figure 1–39 *Notice the ring of gas surrounding the red giant star that has undergone a supernova explosion.*

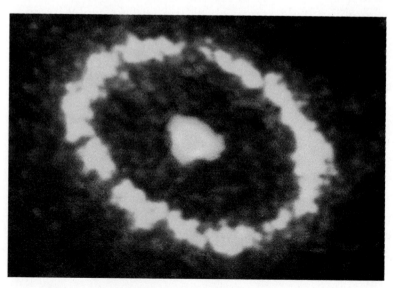

During a supernova explosion, the heat in a star can reach temperatures up to 1,000,000,000°C—that is, one billion degrees Celsius! At these extraordinarily high temperatures, iron atoms within the core fuse together to form new elements. These newly formed elements, along with most of the star's remaining gases, explode into space. The resulting cloud of dust and gases forms a new nebula. The gases in this new nebula contain many elements formed during the supernova. At some point new stars may form within the new nebula.

Most astronomers agree that the nebula from which our sun and its planets formed was the result of a gigantic supernova many billions of years ago. Why do you think astronomers feel the sun and its planets could not have formed in a nebula of only hydrogen and helium gases?

The most famous supernova ever recorded was observed by Chinese astronomers in 1054. The supernova lit the day sky for 23 days and could be seen at night for more than 600 days. Today the remains of this supernova can be seen in the sky as the Crab Nebula. One day, perhaps, new stars will form within the Crab Nebula, and the cycle will begin all over again.

## Neutron Stars

What happens to the remains of the core of a star that has undergone a supernova? Again, the evolution of the star depends on its starting mass. A star that began 1.5 to 4 times as massive as the sun will end up as a neutron star after a supernova. A neutron star is about as massive as the sun but is often less than 16 kilometers in diameter. Such a star is extremely dense. A teaspoon of neutron matter would have a mass of about 100 million tons!

Neutron stars spin very rapidly. As a neutron star spins, it may give off energy in the form of radio waves. Usually the radio waves are given off as pulses of energy. Astronomers can detect these pulses of radio waves if the pulses are directed toward the Earth. Neutron stars that give off pulses of radio waves are called **pulsars.** Thus the end result of a supernova may be a pulsar. And in fact, astronomers have found a pulsar at the center of the Crab Nebula.

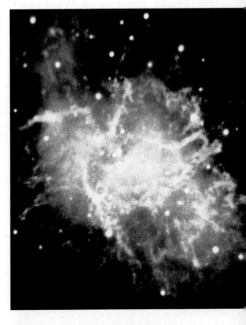

**Figure 1–40** *The Crab Nebula formed from the supernova explosion of a dying star.*

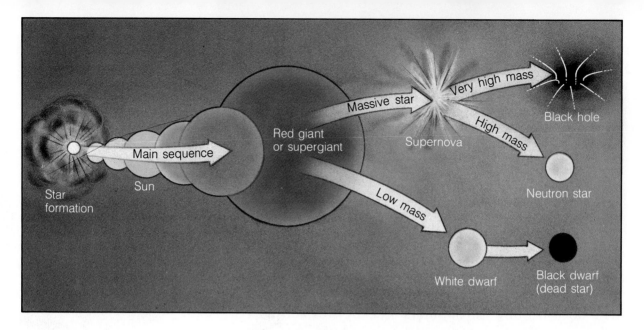

**Figure 1–41** *The fate of a star depends on its mass when it first formed. The sun is a low-mass star that will one day become a white dwarf and finally a dead black dwarf.*

The neutron star in the Crab Nebula pulses at a rate of about 30 times a second. So you can see that a superdense neutron star spins very rapidly.

## Black Holes

Stars with 10 or more times the mass of the sun will have even shorter life spans and a stranger fate than those that wind up as white dwarfs or neutron stars. After a supernova explosion, the core that remains is so massive that, without the energy created by nuclear fusion to support it, the core is swallowed up by its own gravity. The gravity of the core becomes so strong that not even light can escape. The core has become a **black hole.** A black hole swallows matter and energy as if it were a cosmic vacuum cleaner.

If black holes do not allow even light to escape, how can astronomers find them? Actually, it is difficult to detect black holes. But some black holes have a companion star. When the gases from the companion star are pulled into the black hole, the gases are heated. Before the gases are sucked into the black hole and lost forever, they may give off a burst of X-rays. So scientists can detect black holes by the X-rays given off when matter falls into the black hole. See Figure 1–42.

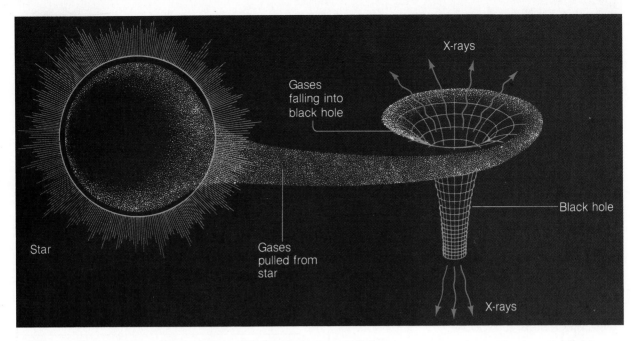

**Figure 1–42** *Some black holes have a companion star. Gases from the companion star are pulled into the black hole. When this occurs, the black hole releases a huge burst of X-rays.*

What happens to matter when it falls into the black hole? Probably the matter is squeezed out of existence, just as a star that becomes a black hole is. But some scientists think strange things may go on inside a black hole. The laws of science may be different within a black hole. Some scientists theorize that black holes are passageways to other parts of the universe, to other universes, or even into time!

## 1–5 Section Review

1. How is the evolution of a star determined by its starting mass?
2. What is the next stage in the sun's evolution?
3. Why are supernovas considered factories for the production of heavy elements?

**Critical Thinking—*Making Inferences***

4. A scientist observes a pulsar in the center of a large nebula. What can the scientist infer about the relationship of the nebula and the pulsar's life cycle?

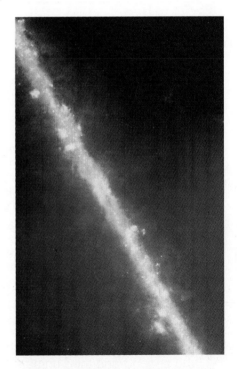

**Figure 1–43** *This photograph of the center of the Milky Way Galaxy was taken with an infrared telescope. Many scientists now believe there is a black hole in the center of the Milky Way.*

# Laboratory Investigation

## Identifying Substances Using a Flame Test

### Problem

How can substances be identified by using a flame test?

### Materials *(per group)*

safety goggles
Bunsen burner
heat-resistant gloves
stainless steel teaspoon
1 unmarked bottle each of sodium chloride, potassium chloride, and lithium chloride

### Procedure

1. Put on the safety goggles. Carefully light the Bunsen burner. **CAUTION:** *If you are not sure how to light a Bunsen burner safely, have your teacher show you the correct procedure.*

2. Put on the heat-resistant gloves.

3. Place the tip of the clean teaspoon in water. Then dip the tip of the spoon into one of the unmarked powders. Make sure that some of the powder sticks to the wet tip.

4. Hold the tip of the spoon in the flame of the Bunsen burner until most of the powder has burned. Observe and record the color of the flame.

5. Repeat steps 3 and 4 using the powder in the second and third unmarked bottles. Observe and record the color of the flame for each powder.

### Observations

| Flame Test | Color of Flame | Name of Substance |
|---|---|---|
| Powder 1 | | |
| Powder 2 | | |
| Powder 3 | | |

### Analysis and Conclusions

1. Sodium chloride burns with a yellow flame. Potassium chloride burns with a purple flame. And lithium chloride burns with a red flame. Using this information, determine the identity of each of the unmarked powders. Record the names of the substances.

2. Why is it important to make sure the spoon is thoroughly cleaned before each flame test? Try the investigation without cleaning the spoon to test your answer.

3. Relate this investigation to the way astronomers study a star's composition.

4. **On Your Own** Predict the color of the flame produced when various combinations of the three powders are used. With your teacher's permission, perform an investigation to test your prediction.

# Study Guide

## Summarizing Key Concepts

### 1–1   A Trip Through the Universe

▲ Nebulae are huge clouds of dust and gas from which new stars are born.

▲ The three types of galaxies are spiral, elliptical, and irregular. Our sun is in the spiral-shaped Milky Way galaxy.

▲ Many stars are found in multiple-star systems.

### 1–2   Formation of the Universe

▲ Every distant galaxy shows a red shift, indicating that the universe is expanding.

▲ Most astronomers agree that the universe began with the big bang.

### 1–3   Characteristics of Stars

▲ Stars range in size from huge supergiants to tiny neutron stars.

▲ The surface temperature of a star can be determined by its color.

▲ Most stars are made up primarily of hydrogen and helium gases.

▲ A star's brightness as observed from Earth is its apparent magnitude. A star's true brightness is its absolute magnitude.

▲ The Hertzsprung-Russell diagram shows the relationship between a star's absolute magnitude and its temperature.

### 1–4   A Special Star: Our Sun

▲ The layers of the sun are the corona, chromosphere, photosphere, and core.

### 1–5   The Evolution of Stars

▲ The main factor that affects the evolution of a star is its starting mass.

## Reviewing Key Terms

*Define each term in a complete sentence.*

### 1–1   A Trip Through the Universe

binary star
constellation
nova
nebula
galaxy
spiral galaxy
elliptical galaxy

### 1–2   Formation of the Universe

spectroscope
spectrum
red shift
Doppler effect
big-bang theory
gravity
quasar

### 1–3   Characteristics of Stars

giant star
supergiant star
white dwarf
neutron star
apparent magnitude
absolute magnitude
Hertzsprung-Russell
  diagram
main-sequence star
parallax
nuclear fusion

### 1–4   A Special Star: Our Sun

corona
chromosphere
photosphere
core
prominence
solar flare
solar wind
sunspot
axis

### 1–5   The Evolution of Stars

protostar
supernova
pulsar
black hole

# Chapter Review

## Content Review

### Multiple Choice

*Choose the letter of the answer that best completes each statement.*

1. Light can be broken up into its character-istic colors by a(an)
   a. optical telescope.  c. spectroscope.
   b. flame test.  d. parallax.
2. The shape of galaxies such as the Milky Way is
   a. elliptical.  c. globular.
   b. irregular.  d. spiral.
3. The most common element in an average star is
   a. hydrogen.  c. helium.
   b. oxygen.  d. carbon.
4. During nuclear fusion, hydrogen atoms are fused into
   a. carbon atoms.  c. iron atoms.
   b. nitrogen atoms.  d. helium atoms.
5. The main factor that shapes the evolution of a star is its
   a. mass.  c. composition.
   b. color.  d. absolute magnitude.

6. The color of a star is an indicator of its
   a. size.  c. surface temperature.
   b. mass.  d. inner temperature.
7. Supermassive stars end up as
   a. main-sequence stars. c. black holes.
   b. neutron stars.  d. white dwarfs.
8. The innermost layer of the sun's atmos-phere is the
   a. corona.  c. chromosphere.
   b. photosphere.  d. core.
9. The most distant objects in the universe are
   a. pulsars.  c. quasars.
   b. neutron stars.  d. binary stars.
10. A star's brightness as seen from Earth is its
    a. absolute magnitude.
    b. average magnitude.
    c. apparent magnitude.
    d. parallax.

### True or False

*If the statement is true, write "true." If it is false, change the underlined word or words to make the statement true.*

1. Our sun is in the <u>Andromeda</u> galaxy.
2. Most stars are <u>main-sequence</u> stars.
3. A solar storm in the form of a huge, bright loop is called a <u>solar flare</u>.
4. Heavy elements are produced in a star during a <u>nova</u> explosion.
5. Most of the core of a red giant is made of <u>helium</u>.
6. In an <u>open universe</u>, all the galaxies will eventually move back to the center of the universe.
7. The <u>blue shift</u> indicates that the universe is expanding.

### Concept Mapping

*Complete the following concept map for Section 1–2. Refer to pages M6–M7 to con-struct a concept map for the entire chapter.*

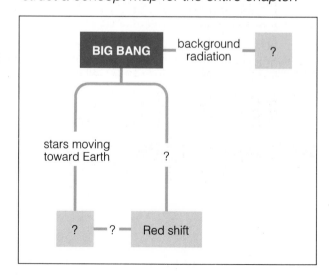

# Concept Mastery

*Discuss each of the following in a brief paragraph.*

1. Compare the evolution of a medium-sized star and a massive star.
2. When you look at the light from distant stars, you are really looking back in time. Explain what this statement means.
3. Describe the Hertzsprung-Russell diagram and the information it provides.
4. What are the different ways astronomers use starlight to study stars?

# Critical Thinking and Problem Solving

*Use the skills you have developed in this chapter to answer each of the following.*

1. **Making predictions** Predict how people in your town would react to a visit by living things from a distant star.
2. **Making comparisons** Compare a pulsar to a lighthouse.
3. **Interpreting diagrams** Examine the spectral lines, or fingerprints, of the elements hydrogen, helium, sodium, and calcium. Compare them with the spectral lines in the spectra labeled X, Y, and Z. Determine which elements produced spectral lines in spectra X, Y, and Z.
4. **Relating cause and effect** Once every three days a small but bright star seems to disappear, only to reappear within six hours. Based on this data, what is causing the small star to disappear?
5. **Using the writing process** You have been on board Earth's first spaceship to another star system for more than six months. Write a letter home to your friends describing your experiences and the wonders you have seen.

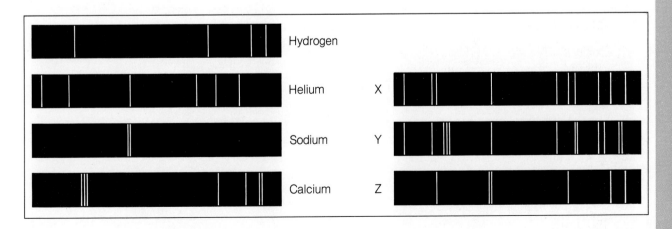

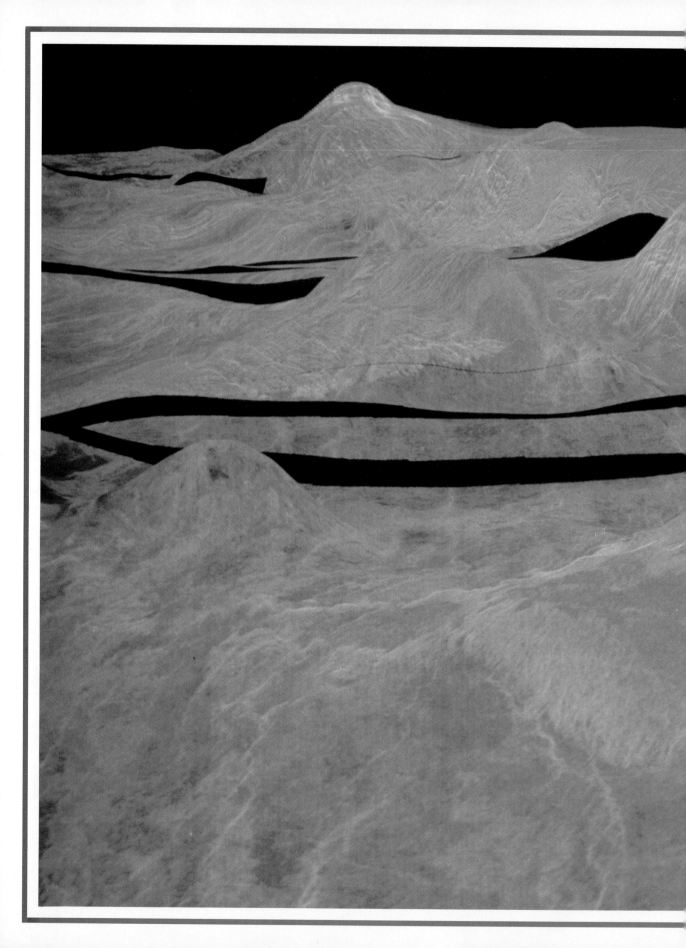

# The Solar System

## Guide for Reading

*After you read the following sections, you will be able to*

**2–1 The Solar System Evolves**
- Describe the nebular theory and the formation of the solar system.

**2–2 Motions of the Planets**
- Identify the shape of a planet's orbit and the factors that contribute to that shape.

**2–3 A Trip Through the Solar System**
- Describe the major characteristics of the planets, moons, asteroids, comets, and meteoroids in the solar system.

**2–4 Exploring the Solar System**
- Describe the principle behind how a reaction engine works.

Planet Venus, next to the sun and moon the brightest object in the night sky, has long been a subject of fascination for Earth-bound observers. One reason for the interest in Venus is that thick clouds blanket the planet, making it impossible to study the Venusian surface from Earth.

How can we study a planet hidden from view? The answer, it turns out, is radar. On August 10, 1991, the United States spacecraft *Magellan* went into orbit above Venus. *Magellan* is a radar-mapping spacecraft. Using complex computers, the craft can produce images of the planet's surface from radar data.

Although it will take many years to analyze the data sent back from *Magellan*, scientists were immediately astounded by some of its discoveries. Pancake-shaped domes that appear to be volcanic in origin are splattered across the surface of Venus. Evidence of lava flows millions of years old crisscross the planet and reveal a highly volcanic planet.

Venus is but one of nine planets in our solar system. One planet—Earth—you call home. In this chapter you will study the other eight planets—and perhaps discover how lucky you are to live on Earth.

### Journal *Activity*

***You and Your World*** Do you already have some idea about what the other planets are like? In your journal, draw and describe one of the planets in the solar system. Then, after you read the chapter, go back and see how close your ideas were to scientific data.

◀ *Scientists constructed this photograph of the Venusian surface using radar data transmitted to Earth from* Magellan.

▶ How does the nebular theory account for the formation of the solar system?

# 2–1 The Solar System Evolves

In Chapter 1 you read about the evolution and life cycle of stars, one of which is our sun. You can also think of the formation of our solar system as a kind of evolution. For, like stars, the solar system changes over time. That is, evolution takes place on a planetary scale. To understand how the solar system formed and has changed, let's take a brief trip back about 5 billion years in time.

In the vast regions between the stars you find yourself in a huge cloud of gas and dust drifting through space. The cloud is cold, colder than anything you can imagine. There is no sun. There are no planets. Slowly moving gases are all that exist. Yet astronomers suggest that our **solar system** formed from this cloud. The solar system includes our sun, its planets, and all the other objects that revolve around the sun.

Many explanations have been proposed to account for the formation of the solar system. But

**Figure 2–1** *The tiny speck of light in this photograph is the distant star Vega. Scientists believe planets may one day form from the ring of gas and dust surrounding the star.*

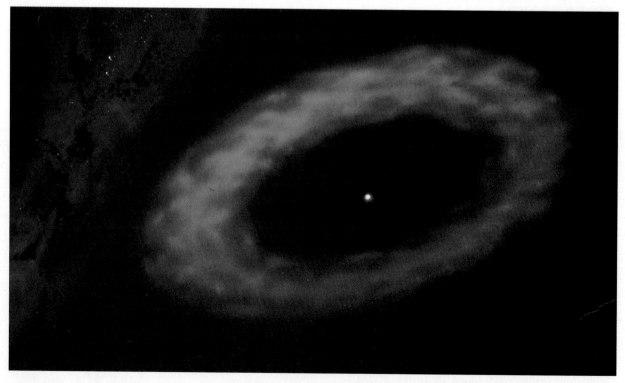

today virtually all astronomers believe in the **nebular theory** of formation. **The nebular theory states that the solar system began as a huge cloud of dust and gas called a nebula, which later condensed to form the sun and its nine planets.** The nebular theory has been revised many times as new data have been gathered. And it will likely be revised many more times. However, studying this theory can reveal much about the mighty forces at work in the formation of our solar system.

## The Sun Forms First

The nebula from which our solar system evolved was composed primarily of hydrogen and helium gases. Yet Earth and the other planets are not made only of these gases. As a matter of fact, the planets contain a wide variety of elements. Where did the elements that make up rivers and mountains, trees and flowers, and even your own body come from? The incredible answer is—from a star!

About 5 billion years ago, according to the nebular theory, a star exploded in a huge supernova. In Chapter 1 you learned that the tremendous heat of a supernova can cause heavy elements to stream into space. These elements rained down on a nearby nebula, seeding it with the chemicals that would become

*A New Planetary System?*

Photographs taken by the *Hubble Space Telescope* have revealed a gaseous disk surrounding a distant star called Beta Pictoris. Astronomers believe the photographs may show that planets orbit Beta Pictoris.

Using reference books in the library, write a report detailing the information about Beta Pictoris that has been discovered by the *Hubble Space Telescope*. Have astronomers found a planet outside our solar system?

**Figure 2–2** *According to the nebular theory, shock waves from a supernova disrupted a nearby nebula. The nebula began to rotate, and gravity pulled more and more matter into a central disk. That central disk became the sun. Clumps of gas and dust around the central disk formed the planets and other objects in the solar system.*

the sun and its planets. At the same time, the shock wave produced by the supernova ripped through the nebula, disrupting the stable gas cloud.

The nebula, which had been slowly spinning, began to collapse. Gravitational forces pulled matter in the nebula toward the center. As the nebula shrank, it spun faster and faster. Gradually, the spinning nebula flattened into a huge disk almost 10 billion kilometers across. At the center of the disk a growing protosun, or new sun, began to take shape.

As the gas cloud continued to collapse toward its center, the protosun grew more and more massive. It became denser as well. In time, perhaps after many millions of years, gravitational forces caused the atoms of hydrogen in the protosun to fuse and form helium. This nuclear fusion gave off energy in the form of heat and light. A star—our sun—was born.

## The Planets Form

Gases and other matter surrounding the newly formed sun continued to spin around the sun. However, the particles of dust and gas were not spread out evenly. Instead, gravity caused them to gather into small clumps of matter. Over long periods of time, some of these clumps came together to form

**Figure 2–3** *This illustration shows the relative sizes of the planets in the solar system. Which planet is the largest?*

larger clumps. The largest clumps became proto-planets, or the early stages of planets.

Protoplanets near the sun became so hot that most of their lightweight gases, such as hydrogen and helium, boiled away. So the inner, hotter proto-planets were left as collections of metals and rocky materials. Today these rocky inner planets are called Mercury, Venus, Earth, and Mars.

The protoplanets farther from the sun were less affected by the sun's heat. They retained their lightweight gases and grew to enormous sizes. Today these "gas giants" are known as Jupiter, Saturn, Uranus, and Neptune.

As the newly formed planets began to cool, smaller clumps of matter formed around them. These smaller clumps became satellites, or moons. Astronomers believe that one of the satellites near Neptune may have broken away from that planet. This satellite became the farthest known planet in the solar system: Pluto. This theory explains why Pluto is similar in composition to many of the icy moons surrounding the outer planets.

Objects other than moons were also forming in the solar system. Between Mars and Jupiter small clumps of matter formed asteroids. These rocklike objects are now found in a region of space between Mars and Jupiter called the asteroid belt. Farther out in space, near the edge of the solar system, other clumps of icy matter formed a huge cloud. Today astronomers believe that this cloud may be the home of comets.

ACTIVITY
WRITING

*An Important Theory*

Using reference materials in your library, write a short report about the theory proposed by the Greek scientist Aristarchus of Samos. What was his theory? Even though the theory was correct, no one believed him. Why?

## 2–1 Section Review

1. Briefly describe how the solar system formed according to the nebular theory.
2. What was the main factor that contributed to the differences between the inner and the outer planets?

**Critical Thinking—*Applying Concepts***
3. Hold a rock in your hand and you are holding stardust. Explain what this statement really means.

## Guide for Reading

*Focus on these questions as you read.*

▶ *What two factors cause planets to move in elliptical orbits around the sun?*

▶ *What is a planet's period of revolution? Period of rotation?*

### *Ellipses*

The orbits of the planets are elliptical. Every ellipse has two fixed points called foci (singular: focus). In any planetary orbit, one of the foci is the sun.

**1.** Stick two thumbtacks into a sheet of stiff paper. The tacks should be about 10 centimeters apart. Wind a 30-centimeter string around the thumbtacks and tie the ends together. Place a sharp pencil inside the string and trace an ellipse. Keep the string tight at all times.

**2.** Repeat step 1, but this time place the thumbtacks 5 centimeters apart.

■ Compare the two ellipses you have drawn. Does the distance between two foci affect the ellipses' shape?

■ Predict what shape you will draw if you remove one of the thumbtacks (foci). Try it.

# 2–2 Motions of the Planets

Long before there were cities or even written language, people looked to the sky for answers to the nature of life and the universe. People used the stars to guide them in traveling and to tell them when to plant crops. Sky watchers knew that although the stars seemed to move across the sky each night, they stayed in the same position relative to one another. For example, a constellation kept the same shape from one night to the next.

In time, however, people who carefully observed the night sky discovered something strange. Some of the "stars" seemed to wander among the other stars. The Greeks called these objects planets, or wanderers. But how were the planets related to Earth and the sun? In what ways did the planets move?

## Earth at the Center?

In the second century AD, the Greek scientist Ptolemy proposed a theory that placed Earth at the center of the universe. Ptolemy also thought that all objects in the sky traveled in **orbits** around an unmoving Earth. An orbit is the path one object takes when moving around another object in space.

In addition, Ptolemy believed that the universe was perfect, unchangeable, and divine. Because the circle was considered the most perfect of all forms, Ptolemy assumed that all objects in space moved in perfectly circular orbits around Earth. The first major challenge to Ptolemy's theory did not come for about 1400 years.

## Sun at the Center?

Between 1500 and 1530 the Polish astronomer Nicolaus Copernicus developed a new theory about the solar system. Copernicus became convinced that Earth and the other planets actually revolved, or traveled in orbits, around the sun.

Based on his theory, Copernicus drew several conclusions regarding the motions of the planets. For one thing, he reasoned that all the planets revolved around the sun in the same direction.

Copernicus also suggested that each planet took a different amount of time to revolve around the sun.

Although Copernicus correctly described many of the movements of the planets, he was wrong about one important concept. Like Ptolemy before him, Copernicus believed that the orbits of the planets were perfect circles.

## Elliptical Orbits

The sixteenth-century German mathematician and astronomer Johannes Kepler supported the theory of Copernicus that planets revolve around the sun. But he discovered something new. After a long and careful analysis of observations made by earlier astronomers, Kepler realized that the planets do not orbit in perfect circles. Instead, each planet moves around the sun in an ellipse, or oval orbit. An oval orbit is approximately egg-shaped.

Today astronomers know that Kepler was correct. Each planet travels in a counterclockwise elliptical

**Figure 2–4** *All the planets revolve around the sun in elliptical orbits. Note that at some points, the orbit of Pluto falls inside that of Neptune. In fact, from 1979 to the year 2000 Neptune will be farther from the sun than Pluto.*

# ACTIVITY

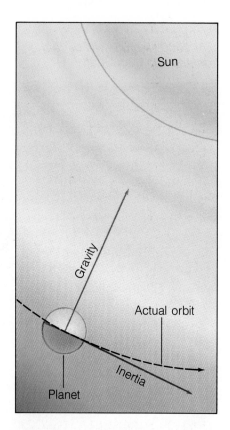

**Figure 2–5** *Inertia makes a planet tend to travel in a straight line (blue arrow). But gravity pulls the planet toward the sun (red arrow). What is the effect of the combined action of inertia and gravity?*

orbit around the sun. Naturally, the planets closest to the sun travel the shortest distance. They complete one orbit around the sun in the shortest amount of time. The more-distant planets travel a longer distance and take a longer time to complete one orbit around the sun. Which planet takes the longest time to complete one orbit?

## Inertia and Gravity

Although Kepler correctly explained the shape of the planets' orbits, he could not explain why the planets stayed in orbit around the sun instead of shooting off into space. In the seventeenth century, the English scientist Sir Isaac Newton provided the answer to that puzzling question.

Isaac Newton began his explanation using the law of inertia. This law states that an object's motion will not change unless that object is acted on by an outside force. According to the law of inertia, a moving object will not change speed or direction unless an outside force causes a change in its motion.

Newton hypothesized that planets, like all other objects, should move in a straight line unless some force causes them to change their motion. But if planets did move in a straight line, they would sail off into space, never to be seen again. Newton realized that some force must be acting on the planets, tugging them into elliptical orbits. That force, he reasoned, is the sun's gravitational pull.

**According to Newton, a planet's motion around the sun is the result of two factors: inertia and gravity.** Inertia causes the planet to move in a straight line. Gravity pulls the planet toward the sun. When these two factors combine, the planet moves in an elliptical orbit. See Figure 2–5.

## Period of Revolution

Another way to say a planet orbits the sun is to say it revolves around the sun. The time it takes a planet to make one revolution around the sun is called its **period of revolution.** A planet's period of revolution is called a year on that planet. For example, Mercury—the planet closest to the sun—takes about 88 Earth-days to revolve once around the sun.

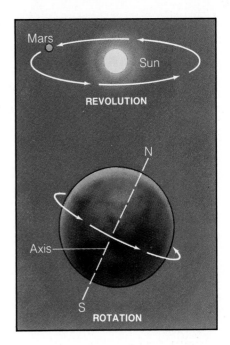

**Figure 2–6** *Mars, like all planets, rotates on its axis while it revolves around the sun. The time it takes to rotate once is called its day. What is the time it takes to make one revolution called?*

REVOLUTION

N

Axis

S

ROTATION

So a year on Mercury is about 88 Earth-days long. Pluto—normally the planet most distant from the sun—takes about 248 Earth-years to revolve around the sun. So a year on Pluto is about 248 Earth-years long.

## Period of Rotation

Aside from revolving around the sun, planets have another kind of motion. All planets spin, or rotate, on their axes. The axis is an imaginary line drawn through the center of the planet. The time it takes a planet to make one rotation on its axis is called its **period of rotation.**

The Earth takes about 24 hours to rotate once on its axis. Does that number seem familiar to you? There are 24 hours in an Earth-day. So the time it takes a planet to go through one period of rotation is called a day on that planet.

Mercury takes almost 59 Earth-days to rotate once on its axis. A day on Mercury, then, is almost 59 Earth-days long. Pluto takes just over 6 Earth-days to rotate once on its axis. So a day on Pluto is a little more than 6 Earth-days long.

## 2–2 Section Review

1. What two factors cause planets to move in elliptical orbits?
2. Describe the two types of planetary motion.
3. Compare the contributions of Ptolemy, Copernicus, Kepler, and Newton to our understanding of planetary motion.

**Connection—*You and Your World***
4. Many stock-car races are run on a track called a tri-oval. Describe or draw the shape of a tri-oval racetrack.

# ACTIVITY
## DOING

*Investigating Inertia*

Inertia is the tendency of an unmoving object to remain in place, or of a moving object to continue in a straight line. Using a toy truck and a rubber band, observe inertia.

**1.** First, attach the rubber band to the front of the truck. Then fill the truck with small rocks or any heavy material.

**2.** Pull on the rubber band until the truck starts to move. Make a note of how much the rubber band must stretch to get the truck moving.

**3.** Continue pulling the truck with the rubber band. Does the rubber band stretch more or less once the truck is moving? Explain.

## Guide for Reading

*Focus on this question as you read.*

▶ What are the major characteristics of the planets and other objects in the solar system?

# 2–3 A Trip Through the Solar System

As you have read, people have wondered about the planets for many centuries. But it was not until recently that spacecraft were sent to examine the planets in detail. From spacecraft observations, one thing about the planets has become clear. **The nine planets of the solar system have a wide variety of surface and atmospheric features.** Figure 2–11 on pages 68–69 presents data about the planets gained from spacecraft and from observations from Earth.

To really appreciate the objects in our solar system, you would have to travel to each of them as spacecraft do. One day you may be able to do this, but for now you will have to use your imagination. So climb aboard your imaginary spaceship and buckle up your safety belt for a journey through the solar system.

## Mercury—Faster Than a Speeding Bullet

Your first stop is a rocky world that has almost no atmosphere. A blazing sun that appears nine times as large as it does from Earth rises in the morning sky. The sun appears huge because Mercury is the closest planet to the sun. Mercury also moves more swiftly around the sun than does any other planet. This tiny world races around the sun at 48 kilometers a second, taking only 88 Earth-days to complete one revolution (one year). This fact explains why the planet was named after the speedy messenger of the Roman gods.

As you approach Mercury, your view is far better than is the view astronomers get from Earth. Mercury is so close to the blinding light of the sun that astronomers on Earth rarely get a good look at it. However, in 1975 the United States spacecraft *Mariner 10* flew past Mercury and provided scientists with their first close look at the planet.

*Mariner 10* found a heavily crater-covered world. The craters were scooped out billions of years ago

**Figure 2–7** *These photographs of Mercury were taken by* Mariner 10. *The craters were scooped out of the surface billions of years ago. Why have they remained unchanged over all that time?*

by the impact of pieces of material striking the surface of the planet. Because Mercury has almost no atmosphere, it has no weather. Since there is no rain, snow, or wind to help wear down the craters and carry away the soil particles, the craters of Mercury appear the same as when they were created. As a result, Mercury has changed very little for the past few billion years.

Photographs from *Mariner 10* also revealed long, steep cliffs on Mercury. Some of the cliffs cut across the planet for hundreds of kilometers. There are also vast plains. These plains were probably formed by lava flowing from volcanoes that erupted billions of years ago. There is no evidence today of active volcanoes on Mercury.

As you read, Mercury rotates on its axis very slowly, taking about 59 Earth-days for one complete rotation. In fact, Mercury rotates three times about its axis for every two revolutions around the sun. The combined effect of these two motions produces a sunrise every 175 Earth-days. So the daytime side of the planet has lots of time to heat up, while the nighttime side has plenty of time to cool off. This long period of rotation causes temperatures on Mercury to range from a lead-melting 427°C during the day to −170°C at night. Mercury is, therefore, one of the hottest and one of the coldest planets in the solar system.

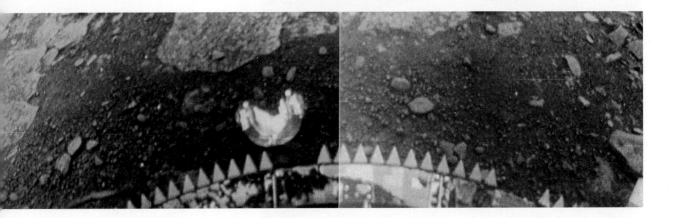

**Figure 2–9** *This photograph of the Venusian surface was taken by a Soviet* Venera *spacecraft, a portion of which can be seen at the bottom of the photograph. Soon after taking this photograph, the spacecraft went silent—a harsh reminder of the extreme conditions on Venus.*

After taking off and leaving Mercury behind, you speed deeper into the solar system. Your next stop is the second planet from the sun.

## Venus—Greenhouse in the Sky

Venus, the next stop on your imaginary journey, was named for the Roman goddess of beauty and love. Venus has about the same diameter, mass, and density as Earth does. For these reasons, astronomers once called Venus Earth's twin. People even imagined that Venus, like Earth, might be covered with vast oceans and tropical forests. For many years no one was sure whether this was true or not.

The uncertainty about Venus was due to its thick cloud cover, which has covered Venus for 400 to 800 million years. Clouds on Venus are more than five times as dense as are clouds on Earth. From Earth, astronomers can see only the yellowish Venusian clouds.

In recent years, however, data from spacecraft have slowly revealed the surface of Venus to astronomers. In 1975, two Soviet spacecraft (*Venera 9* and *Venera 10*) landed on Venus. The spacecraft were not able to withstand the harsh conditions and functioned for only a few hours. But before they failed, they were able to send back the first photographs of the Venusian surface. More recently, two United States spacecraft, *Pioneer Venus Orbiter* and *Magellan,* were placed in orbit around Venus. Radar instruments were able to penetrate the thick cloud cover and map much of the Venusian surface. The story that follows is based on information gathered from such probes.

As you approach Venus, your instruments detect winds of more than 350 kilometers per hour pushing the upper cloud layers around the planet. When you descend into the yellow clouds, you discover that they are not made of water vapor, as clouds on Earth are. These clouds consist of droplets of sulfuric acid and carbon dioxide. As you descend farther into this hostile atmosphere, the temperature and pressure rise rapidly. Sulfuric-acid rain falls through the cloud layers but evaporates in midair, never reaching the surface. Bolts of lightning flash near your ship.

Finally, you reach the surface. The atmosphere near the surface contains mainly carbon dioxide and is bathed in an eerie orange glow. Temperatures climb to 480°C, even hotter than on the surface of Mercury. No water has been found on Venus. The thick atmosphere bears down on you with a pressure 91 times greater than the atmospheric pressure at sea level on Earth.

As your craft skims over the surface of Venus, you discover deep canyons, craters, and vast plains. The remains of once-active volcanoes dot the surface, appearing like pancakes or upside-down cereal bowls. Venus also has a few continent-sized highland areas. In the distance you spot mountains as tall as any on Earth. Scientists feel these mountains were formed by ancient Venusian volcanoes. These volcanoes were likely the source of the thick atmosphere that covers Venus. You also notice a huge crack, or channel, in the surface. This channel runs for almost 7000 kilometers, longer than the Nile River and deeper than the Grand Canyon back on Earth.

From the surface, the cloud cover completely hides your view of the sun. But if you could see the sun, you would see something that would be a totally new experience. The sun would slowly rise in the west and later set in the east. The sun follows this pattern because, unlike Earth, Venus rotates from east to west. Astronomers call this reverse motion **retrograde rotation.** Another unusual aspect of Venus is that it rotates once on its axis every 243 Earth-days. However, Venus takes only 224 Earth-days to revolve once around the sun. A Venusian day, then, is actually longer than a Venusian year.

**Figure 2–10** *As you can see in this photograph, Venus is a planet covered by thick clouds.*

## THE SOLAR SYSTEM

| Name | Average Distance From Sun (millions of km) | Diameter (km) | Period of Revolution in Earth-time Days | Years | Period of Rotation Days | Hours | Number of Moons |
|------|------|------|------|------|------|------|------|
| Mercury | 58 | 4880 | 88 | — | 58 | 16 | 0 |
| Venus | 108 | 12,104 | 225 | — | 243 Retrograde | — | 0 |
| Earth | 150 | 12,756 | 365 | — | — | 24 | 1 |
| Mars | 228 | 6794 | — | 1.88 | — | 24.5 (about) | 2 |
| Jupiter | 778 | 142,700 | — | 11.86 | — | 10 (about) | 16 |
| Saturn | 1427 | 120,000 | — | 29.46 | — | 10.5 (about) | 23? |
| Uranus | 2869 | 50,800 | — | 84.01 | — Retrograde | 16.8 (about) | 15 |
| Neptune | 4486 | 48,600 | — | 164.8 | — | 16 | 8 |
| Pluto | 5890 | 2300 | — | 247.7 | 6 | 9.5 | 1 |

**Figure 2–11** *This chart shows the most current information known about the planets. Which planets show retrograde rotation?*

Your stay on Venus is almost over. By now you have discovered that Venus is certainly not the twin of Earth. But why is Venus, the closest planet to Earth, so vastly different from our world? Why is it such a dry, hot world? Billions of years ago, when the solar system was still forming, the sun was much cooler than it is today. In those early days, Venus may have been covered with planet-wide oceans. In fact, the remains of coastlines and sea beds can still be detected today. Then, as the sun grew hotter, water began to evaporate into the atmosphere. This water vapor helped to create a heat-trapping process

| Temperature Extremes (°C) High | Low | Orbital Velocity (km/sec) | Atmosphere | Main Characteristics |
|---|---|---|---|---|
| 427 | −170 | 47.8 | Hydrogen, helium, sodium | Rocky, cratered surface; steep cliffs; extremely thin atmosphere |
| 480 | −33 | 35.0 | Carbon dioxide | Thick cloud cover, greenhouse effect, vast plains, high mountains |
| 58 | −90 | 29.8 | Nitrogen, oxygen | Liquid water, life |
| −31 | −130 | 24.2 | Carbon dioxide, nitrogen, argon, oxygen, water vapor | Polar icecaps, pink sky, rust-colored surface, dominant volcanoes, surface channels |
| 29,700 | −95 | 13.1 | Hydrogen, helium, methane, ammonia | Great red spot, thin ring, huge magnetosphere, rocky core surrounded by liquid-hydrogen ocean |
| ? | −180 | 9.7 | Hydrogen, helium, methane, ammonia | Many rings and ringlets, Titan only moon with substantial atmosphere |
| ? | −220 | 6.8 | Hydrogen, helium, methane | Rotates on side, 9 dark mostly narrow rings of methane ice, worldwide ocean of superheated water |
| ? | −220 | 5.4 | Hydrogen, helium, methane | Unusual satellite rotation, 4 rings, great dark spot, rocky core surrounded by slush of water and frozen methane |
| ? | −230 | 4.7 | Methane | Smallest planet, possibly a double planet |

called the **greenhouse effect.** The greenhouse effect occurs when heat becomes trapped beneath the clouds.

As the temperature rose further, the oceans evaporated completely. However, even after all the water was gone from Venus, the greenhouse effect continued. The atmosphere of Venus is mainly carbon dioxide. The carbon dioxide, like the water vapor before it, traps heat and produces a greenhouse effect. So today, even during the long nights on Venus, the dark side of the planet remains about as hot as the bright side.

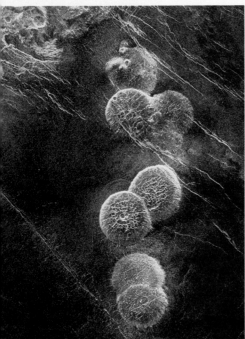

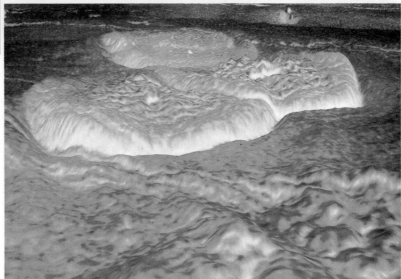

**Figure 2–12** *These remarkable photographs of the Venusian surface were developed using radar data from the* Magellan *spacecraft. They show a planet dominated by volcanoes and deep valleys. The pancake-shaped structures are the domes of volcanoes.*

# ACTIVITY
## DOING

*Build a Greenhouse*

How did the greenhouse effect get its name? Fill two containers with potting soil. Place a thermometer on the surface of the soil in each container. Cover one container with a sheet of glass. Put both containers in a sunny window. Observe what happens to the temperature in each container.

If the term greenhouse effect seems familiar to you, it is probably because scientists warn of a similar problem on Earth. The Earth's atmosphere also acts as a greenhouse. Up until now, this has kept the Earth warm enough for life to evolve and survive. However, the burning of fossil fuels such as coal and oil adds carbon dioxide to the Earth's atmosphere. Scientists fear that this increased carbon dioxide may cause a runaway greenhouse effect, much like the one that left Venus dry, hot, and barren. What are some ways to prevent a runaway greenhouse effect from happening on Earth?

Even on an imaginary trip, the harsh conditions on Venus make you uncomfortable. So you decide to continue your journey. For now, however, you will skip the third planet, Earth.

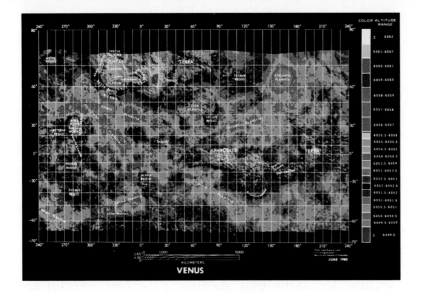

**Figure 2–13** *This map of the Venusian surface was produced using radar data from a spacecraft orbiting Venus.*

## Mars—The Rusty Planet

Your imaginary ship is now approaching Mars, the fourth planet from the sun. As you reach Mars, the first thing you notice is its reddish color. In ancient times, this color reminded people of blood, and they thought of Mars as a warrior planet. Today it still bears the name of the Roman god of war. Appropriately, the two tiny moons that circle Mars are called Phobos and Deimos, from the Greek words for fear and terror.

In late July 1976, a spacecraft landed successfully on Mars. This was not an easy task, for the surface of Mars is rocky and heavily cratered. The ship, named *Viking 1* after early explorers on Earth, was soon followed by a second ship, *Viking 2*. Both quickly began to send back detailed photographs of the Martian surface. Another giant step in the exploration of the solar system had been taken.

# Activity Bank

Rusty Nails, p.149

**Figure 2–14** *Notice the characteristic red color of the Martian soil in this photograph taken by a Viking spacecraft that landed on Mars. What causes the soil on Mars to appear red?*

**Figure 2–15** *You can actually see the thin Martian atmosphere over the horizon in this photograph taken by* Viking 1.

One of the most important tasks of the Viking spacecraft was to analyze Martian soil. To do so, a robot arm scooped up some of the soil and placed it in the on-board laboratory. Tests revealed that Martian soil is similar to Earth's soil in many ways. But there are differences. For centuries Mars had been known as the red planet. Soil tests showed why this is so. Martian soil is coated with a reddish compound called iron oxide. Perhaps you know iron oxide by its more common name—rust!

The Viking spacecraft also tested the soil for signs of life. Although the tests did not reveal any signs of life or life processes, the data did not rule out the possibility that life may once have existed on Mars.

The Viking spacecraft as well as observations from the Earth aided in the discovery of many other features on Mars. Mars appears to be a planet that has had a very active past. For example, four huge volcanoes are located on Mars. These volcanoes are dormant, or inactive. But large plains covered with lava indicate that the volcanoes were once active. The largest volcano on Mars is *Olympus Mons. Olympus Mons* is wider than the island of Hawaii, and it is almost three times as tall as Mount Everest. In fact, *Olympus Mons* is the largest known volcano in the solar system.

Astronomers now believe that when the Martian volcanoes were active, they poured out both lava and steam. As the steam cooled, it fell as rain. Rushing rivers may have once crossed the Martian surface, gouging out channels that wander across Mars. Today there is no liquid water on Mars. But frozen water can be found in the northern icecap and may also be located under the soil.

**Figure 2–16** *The dead volcano* Olympus Mons *on Mars is the largest volcano ever discovered.*

The northern icecap of Mars is made mostly of frozen water, which never melts. But the southern icecap is mostly frozen carbon dioxide. Much of this icecap melts during the Martian summer. But do not be misled by the word summer. Since Mars has a very thin atmosphere made mostly of carbon dioxide, it does not retain much heat from the sun. So even during the summer, temperatures on Mars are well below 0°C. That, of course, is why water on Mars stays frozen all year round.

Another interesting feature of Mars is an enormous canyon called *Valles Marineris*. The canyon is 240 kilometers wide at one point and 6.5 kilometers deep. If this canyon were on the Earth, it would stretch from California to New York.

Although the atmosphere of Mars is very thin, winds are common. Windstorms sweep across the surface at speeds up to 200 kilometers per hour. These storms stir up so much dust that the sky may turn a dark pink.

As you have read, Mars has two moons called Phobos and Deimos. These rocky, crater-covered moons are much smaller than Earth's moon. The maximum diameter of Phobos is only 25 kilometers. The diameter of Deimos is only 15 kilometers.

Your stay on Mars is just about over—and just in time. Another dust storm has begun to develop.

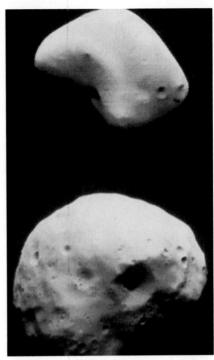

**Figure 2–17** *Both moons of Mars, Phobos and Deimos, are shown in this composite NASA photograph.*

## The Asteroid Belt

Your journey from Mars to Jupiter holds a new kind of danger. Thousands, perhaps hundreds of thousands, of rocks and "flying mountains" lie in your path. These objects are the "minor planets," which sweep around the sun between the orbits of Mars and Jupiter. You have entered the **asteroid belt.**

Asteroids may be made of rocks, metals, or a combination of the two. Most asteroids are small and irregularly shaped. A few, however, are huge. The largest asteroid, Ceres, has a diameter of almost 1000 kilometers. Earth's moon, by comparison, is about 3400 kilometers in diameter.

At one time, astronomers thought that the fragmented objects in the asteroid belt were the remains of a planet that broke apart long ago. However, it now appears that the asteroid belt is made up of clumps of matter that failed to join together to form a planet during the birth of the solar system. Why? Scientists suspect that Jupiter's strong gravitational pull kept the asteroids from coming together.

Not all asteroids are found in the asteroid belt. For example, some asteroids hurtle through space close to the Earth. Fortunately, these "flying mountains" are rarely on collision courses with Earth. However, collisions do occur from time to time. Many impact scars on Earth have been identified. At least seventy of these scars are thought to be the results of asteroids plowing into Earth's crust at tremendous speeds. Many of the craters on the moon and on other planets may also be due to the impact of asteroids.

One theory that has prompted a good deal of scientific debate states that the collision of a huge asteroid some 65 million years ago resulted in changes that led to the extinction of the dinosaurs and almost 90 percent of all other life on Earth at that time. It has been estimated that the force of the asteroid collision may have been some 10,000 times greater than the force that would result if all the nuclear weapons on Earth were exploded at one time!

Although you have had to steer your ship carefully, you have managed to pass safely through the asteroid belt. The giant planet Jupiter looms ahead.

**Figure 2–18** *The asteroid belt is a region located between the orbits of Mars and Jupiter. What is the composition of most asteroids?*

**Figure 2–19** *Many scientists believe that the extinction of the dinosaurs was caused by the collision of an asteroid some 65 million years ago. How might such a collision have caused this to happen?*

## Jupiter—The Planet That Was Almost a Star

The first thing you notice as you approach Jupiter is its size. Our sun contains about 99.8 percent of all the matter in the solar system. Jupiter contains about 70 percent of what is left. A hundred Earths could be strung around Jupiter as if they were a necklace of pearls. Jupiter is so big and bright in the night sky that the Romans named this planet after their king of the gods.

In many ways Jupiter rivals the sun. Like the sun and other stars, Jupiter is made primarily of hydrogen and helium gases. The temperature is cold at the cloud tops but rises considerably beneath the upper cloud layers. At Jupiter's core, scientists believe temperatures may reach 30,000°C, almost five times the surface temperature of the sun. Scientists think that if Jupiter had grown larger during the formation of the solar system, gravitational forces might have caused nuclear fusion to occur and a star to form. So you can think of Jupiter as a planet that was almost a star.

From Earth, all that can be seen of Jupiter's atmosphere is its thick cloud cover. These clouds,

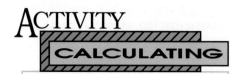

## Activity

### CALCULATING

*Comparing Diameters*

Jupiter has a diameter of 142,800 kilometers. Mercury has a diameter of 4900 kilometers. How many times larger is Jupiter than Mercury?

which appear as bands of color, are made mostly of hydrogen and helium. Other gases, such as ammonia and methane are also found in Jupiter's atmosphere.

As your imaginary ship nears Jupiter, you notice that the clouds are very active. Huge storms swirl across the surface of the atmosphere. These storms can be observed because the colored bands of the clouds are twisted and turned by the strong winds. Perhaps the best-known feature of Jupiter's cloud cover is a giant red spot three times the size of Earth. This Great Red Spot, which is probably a hurricanelike storm, has been observed for more than 100 years. (Scientists estimate it may be well over 20,000 years old.) If it is a storm, it is the longest-lasting storm ever observed in the solar system.

Unlike the other planets you have read about, Jupiter probably has only a small solid core. The clouds become thicker and denser as they get closer to the center of the planet's core. As their density increases, the clouds may change into a giant ocean of liquid hydrogen.

Because of the thick cloud cover, the atmospheric pressure on Jupiter is enormous. In fact, the pressure near the center of the planet is so great that the liquid-hydrogen ocean probably changes into a form of liquid hydrogen that acts like a metal. This liquid metallic layer may surround a rocky core about the size of Earth. The liquid metallic layer is the cause of Jupiter's gigantic magnetic field. The magnetic field, called the **magnetosphere**,

**Figure 2–20** *This photograph of Jupiter was taken by a Voyager spacecraft. The ring has been added by an artist. Can you find the giant red spot on Jupiter?*

**Figure 2–21** *In this composite photograph, Jupiter and its four largest moons are shown. What are the four largest moons called?*

stretches for millions of kilometers beyond the planet. Jupiter's magnetosphere is the largest single structure in the solar system. Jupiter is unusual in other ways. For example, it gives off more heat than it receives from the sun.

In 1979, two Voyager spacecraft (*Voyager 1* and *Voyager 2*) flew past Jupiter. These spacecraft took thousands of photographs of the gas giant. From these photographs, astronomers discovered a thin ring circling Jupiter. They also discovered gigantic bolts of lightning in the atmosphere and mysterious shimmering sheets of light in the sky.

In 1610, the scientist Galileo Galilei observed four moons orbiting Jupiter. Today these moons are known as the Galilean satellites. And although at least sixteen moons have now been found orbiting Jupiter, the four largest and most interesting are the moons discovered by Galileo more than 300 years ago.

**Io** The innermost of Jupiter's large moons is Io. Io is perhaps the most dramatic object in the solar system. The moon seems painted in brilliant orange, yellow, and red hues, which are due mainly to the high sulfur content of Io's surface. This mix of colors

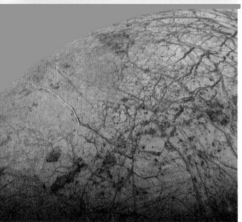

**Figure 2–22** *Io, the innermost of Jupiter's largest moons, has a surface that is constantly changed by volcanic eruptions (top). Europa shows tan streaks that may be shallow valleys (bottom).*

prompted one scientist to compare the colorful moon to a pepperoni pizza. Scientists originally assumed Io's surface would be heavily cratered. They expected to see the scars of impacts with large objects that occurred over the past few billion years. Instead, the scientists found a young, active surface. The surface looked so young because it was constantly being covered by new material from Io's active volcanoes. Today scientists consider Io the most geologically active object in the solar system.

**EUROPA** Next out from Jupiter is Europa. Europa is an ice-covered world slightly smaller than Io. Europa has the brightest, whitest, and smoothest surface of any object astronomers have observed in the solar system. It has been described as a giant "billiard ball in the sky." Some of the photographs from the Voyager spacecrafts indicate that Europa may have a volcano that spews out water and ammonia ice. Such a volcano would be far different from those on Earth, which spew out molten rock.

**GANYMEDE** Beyond Europa is Ganymede, Jupiter's largest moon. Ganymede, in fact, is the largest moon in the solar system. It is larger even than the planet Mercury. Ganymede is an icy world, about half rock and half water ice. It has some smooth regions, but it also has craters.

Some regions on Ganymede look as though they have been shaken by "earthquakes." Pieces of the moon's surface look as though they have cracked and slipped past one another. If that is what happened, Ganymede is the first object in the solar system besides Earth and its moon that is known to have earthquakes.

**CALLISTO** Your next stop is Callisto, the most heavily cratered object in the solar system. Although they are very small, Callisto's craters cover almost every part of this moon's surface. Scientists estimate that it would have taken several billion years of impacts to punch out all the craters of Callisto. Therefore, the surface of Callisto, which is mainly rock and ice, appears much as it did billions of years ago. This further suggests that Callisto is, and has been, a very quiet world. If volcanoes such as those on Io existed on Callisto, they would have filled in many of the craters.

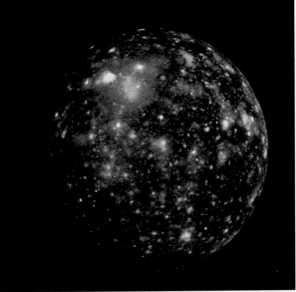

There is much that can still be learned about the moons of Jupiter. Now, however, it is time to journey to the gas giant that is the second largest planet in the solar system.

Figure 2–23 *Ganymede is a moon that is half covered by ice and half covered by rock (left). The bright spots on the surface of Callisto are craters billions of years old (right).*

## Saturn—A World of Many Rings

As you approach Saturn, you notice that it is surrounded by a series of magnificent rings. Saturn's rings were discovered by Galileo, and it was the first planet found to have rings. Many astronomers consider Saturn's rings to be the most beautiful sight in the solar system—so enjoy your view.

The rings of Saturn are made mainly of icy particles ranging in size from one thousandth of a millimeter to almost 100 kilometers in diameter. When observed through a telescope, Saturn appears to have three main rings. However, photographs taken by Voyager spacecraft showed that Saturn's ring system is far more complex than we could ever tell from Earth. Voyager revealed that Saturn has at least seven major rings, lettered from A to G. The outer edge of the most distant ring is almost 300,000 kilometers from Saturn. In addition, the main rings are made up of tens of thousands of ringlets that weave in and out of the main rings.

While Saturn's rings are its most spectacular feature, the planet is also interesting in other ways. Like Jupiter, Saturn spins rapidly on its axis and is made mainly of hydrogen and helium gases. Because

ACTIVITY

DISCOVERING

*Orbital Velocities*

Using the data in Figure 2–11, make a graph of the orbital velocities of the planets in our solar system. Plot orbital velocity on the vertical axis and the nine planets in order on the horizontal axis.

■ What conclusions can you draw from the curve on your graph?

**Figure 2–24** *Saturn's ring system may well be the most beautiful sight in the solar system. What is the composition of Saturn's rings?*

Saturn spins so fast, it is flattened at the poles, and it bulges at the equator. Near the equator, winds speed around Saturn at about 1800 kilometers per hour. This superfast jet stream is four times as quick as the fastest winds of Jupiter. Also like Jupiter, Saturn has violent storms. *Voyager 2* detected one enormous lightning storm that lasted more than ten months.

Saturn's clouds, like Jupiter's, form colored bands around the planet. Light-colored bands alternate with darker bands. There is even a reddish-orange oval feature in Saturn's southern hemisphere, a smaller version of Jupiter's Great Red Spot. Saturn is colder than Jupiter, yet it gives off almost three times as much energy as it gets from the sun. Saturn also has a huge magnetic field, second in size only to Jupiter's magnetosphere. Scientists suspect that the core of Saturn may also be similar to Jupiter's small, inner core.

Another unusual feature of Saturn is its very low density. In fact, Saturn is the least dense planet in the solar system. If all the planets could be placed in a giant ocean, Saturn would be the only planet to float on water.

As you fly by Saturn you must be careful not to collide with one of its moons. Saturn has more moons than any other planet. So far, twenty-one—and possibly two more—moons have been found orbiting Saturn. The largest of Saturn's moons is Titan. Only Ganymede is larger than Titan. However, size is not the only thing that makes Titan an unusual moon. Titan has a substantial atmosphere. The atmosphere is mainly nitrogen, but also contains methane, hydrogen cyanide, carbon monoxide, carbon dioxide, and other gases. The combination of these gases gives Titan a hazy orange glow. Many of these gases are deadly to life on Earth. But it is interesting to note that before life formed, the Earth had an atmosphere very similar to that on Titan. Some scientists have speculated that living things could evolve in the atmosphere of Titan, although no life has been detected.

**Figure 2–25** *Notice the haze, indicating an atmosphere, surrounding Saturn's moon Titan. What is the composition of Titan's atmosphere?*

**Figure 2–26** *In this composite photograph you can see Saturn surrounded by six of its moons. The large moon in front is Dione. The other moons (clockwise) are Enceladus, Rhea, Titan, Mimas, and Tethys.*

## Uranus—A Planet on Its Side

You have now traveled almost 1.5 billion kilometers on your imaginary trip through the solar system. But you still have quite a distance to go before you reach the seventh planet, Uranus. Named for the father of Saturn in Roman mythology, Uranus was discovered in 1781 by the English astronomer Sir William Herschel. With Herschel's discovery, the size of the known solar system doubled, for Uranus is almost twice as far from the sun as is Saturn. See Figure 2–31 on pages 84–85 for the relative distances of the planets from the sun.

As you approach Uranus, you notice immediately that it is a gas giant, much like Jupiter and Saturn. Uranus is covered by a thick atmosphere made of hydrogen, helium, and methane. The clouds of Uranus do not have bands, but rather the entire

**Figure 2–27** *This photograph of the gas giant Uranus was taken by a Voyager spacecraft as it passed the distant planet.*

**Figure 2–28** *Notice the craters on Miranda, a moon of Uranus.*

**Figure 2–29** *This composite photograph shows a portion of Uranus with Miranda, one of its moons, in the foreground.*

planet is tinged with a greenish-blue color. Temperatures at the top of the clouds may dip to as low as −220°C.

Data from *Voyager 2* provide strong evidence that Uranus is covered by an ocean of superheated water that may have formed from melted comets. Because of the extreme pressure from an atmosphere 11,000 kilometers thick, the superheated water does not boil. This worldwide ocean is about 8000 kilometers thick and encloses a rocky, molten core about the size of Earth.

The axis on which Uranus rotates is one of the most unusual features of this gas giant. The axis of Uranus is tilted at an angle of about 90°. So Uranus seems to be tipped completely on its side. Uranus has nine known rings. But unlike the rings of Saturn, these rings are dark and probably made of methane ice. Because of the tilt of the axis of Uranus, the rings appear to circle the planet from top to bottom.

The *Voyager 2* flyby confirmed the fact that Uranus has fifteen moons, ranging in diameter from 32 to 1625 kilometers. Some of the more interesting

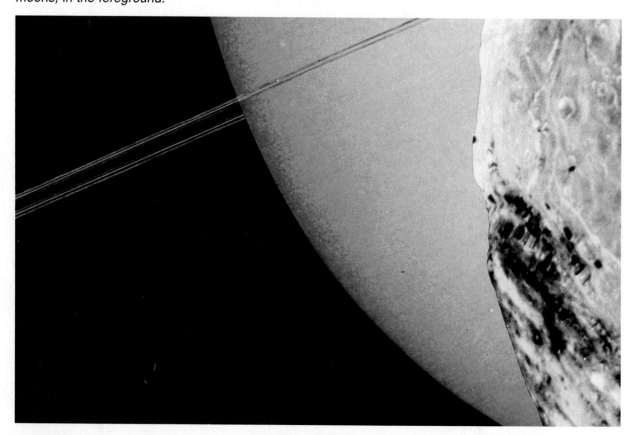

moons include Miranda and Ariel. Both Miranda and Ariel are geologically active, and their surfaces show many fault lines. (Faults are places where landmasses collide and are the cause of earthquakes on Earth.)

## Neptune—The Mathematician's Planet

Soon after the discovery of Uranus, astronomers found that the blue-green planet was not behaving as expected. Uranus was not following exactly the orbital path that had been carefully calculated for it by taking into account the gravitational pull of the other planets. Astronomers decided that there must be another object beyond Uranus. The gravitational pull from this distant object in space, it was assumed, was affecting the orbit of Uranus.

In 1845, a young English astronomer John Couch Adams calculated where such an object should be. For the most part his results were ignored. Meanwhile, in France, Urbain Jean Joseph Leverrier also calculated the location of this new planet. Leverrier's calculations were also largely ignored. However, one scientist, Johann Galle at Germany's Berlin Observatory, took Leverrier seriously. Galle immediately began searching for the unknown object. Before his first night of observation was over, he had discovered a new planet. It was located exactly where both Adams and Leverrier had predicted the mysterious object would be. And it is that mysterious object you will visit next on your tour of the solar system.

The new planet, a giant bluish world, was named Neptune, for the Roman god of the sea. Neptune and Uranus are often called the twin giants. They are about the same size, mass, and temperature. Neptune also glows with a blue-green color.

Like Uranus, Neptune is covered by a thick cloud cover. Huge clouds of methane float in an atmosphere of hydrogen and helium. Temperatures at the cloud tops may dip to a chilly -220°C. Neptune's surface is probably an ocean of water and liquid methane, covering a rocky core.

Data from *Voyager 2* confirmed that Neptune has five rings. These rings are made of dust particles that may have formed when meteorites crashed into Neptune's moons millions of years ago.

# ACTIVITY

## THINKING

*Outer Planetary Weather*

Saturn and Neptune are the windiest planets in the solar system, and Jupiter is the stormiest. Using information from the chapter, create a weather forecast for the planets Jupiter, Saturn, and Neptune. Assume you are a local weather forecaster providing a traveler's forecast for people on Earth who will be journeying to these distant planets.

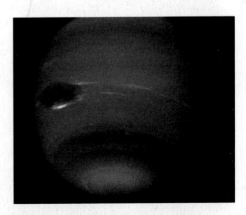

**Figure 2–30** *After passing by Uranus, Voyager continued on to Neptune, where it took this photograph of the gas giant. Why are Uranus and Neptune called the twin giants?*

**Figure 2–31** *This illustration shows the relative distances of the planets from the sun, not their sizes. The closest planet, Mercury, averages about 58 million kilometers from the sun. The farthest planet, Pluto, averages about 5900 million kilometers from the sun.*

Neptune also has at least eight moons. The most interesting moon is Triton, the fourth largest moon in the solar system. Triton appears to be an icy world covered with frozen methane. Like Titan, Triton has an atmosphere. Triton is an unusual moon because it orbits Neptune in a backward, or retrograde, direction. This fact has led some astronomers to conclude that Triton is not an original moon of Neptune. Instead, it may be an object captured by Neptune's gravity.

Your journey through the solar system is not quite over. Now it is time to travel to the only planet that was not discovered until this century.

**Figure 2–32** *Triton, Neptune's largest moon, is a world covered with frozen methane.*

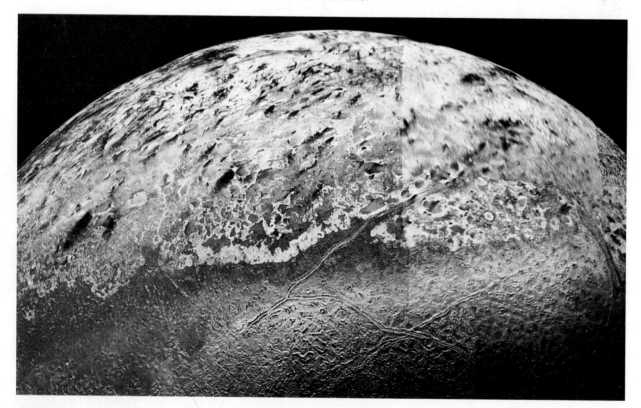

## Pluto—A Double Planet

Neptune's discovery helped to explain some of the unexpected changes in the orbit of Uranus. But it did not account for all the changes. To complicate matters, the newly discovered Neptune did not orbit the sun as predicted either. In the early 1900s, astronomer Percival Lowell attempted to explain the mystery. He suggested that there was another planet whose gravity was pulling on both Neptune and Uranus.

In 1930, after an intense search, a young astronomy assistant named Clyde Tombaugh found the ninth planet near where Lowell had predicted it would be. The planet was named Pluto, for the Roman god of the underworld. However, the discovery of Pluto still did not solve the riddle of the strange orbits of Uranus and Neptune.

Lowell had calculated the position of a world that he thought was huge—a world massive enough to pull the two gas giants Uranus and Neptune out of their expected orbits. But, as it turns out, Pluto is much too small to have any real effect on either of these giant planets. In fact, Pluto is the smallest and least massive planet in the solar system.

Pluto is little more than a moon-sized object and may be an escaped moon of Neptune. It appears to be made mainly of various ices, primarily methane ice. Although the methane is frozen on the dark side of Pluto, it seems likely that some of the methane on the part of the planet facing the sun may have evaporated and formed a thin, pink atmosphere. If so, Pluto would be the only planet with an atmosphere on its sunny side and no atmosphere on its dark side.

As your imaginary ship approaches Pluto, you notice something that remained hidden to Earthbound astronomers for forty-eight years after

## ACTIVITY DOING

*Planetary Sizes*

Examine the diameters of the planets in our solar system as shown in Figure 2–11. Using art materials and measuring tools, illustrate visually the relative sizes of the planets. Keep in mind that everything in your model must be done to the same scale.

**Figure 2–33** *This NASA illustration shows Pluto and its moon Charon. Why do some people call Pluto a double planet?*

the discovery of Pluto. In 1978, astronomer James Christy was studying photographs of Pluto when he noticed some that appeared to be defective. In the "defective" photograph, Pluto seemed to have developed a bump. Looking more closely, Christy realized that the bump was not part of the planet. It was a moon. He named the moon Charon, after the mythological boatman who ferried the souls of the dead into the underworld. Charon is about half the size of Pluto. Because of this closeness in size, astronomers consider Pluto and Charon to be a double planet.

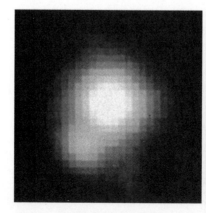

**Figure 2–34** *The photograph on the left shows Pluto and its moon as seen from Earth. The photograph on the right was taken by the Hubble Space Telescope.*

## Planet X—The Tenth Planet?

As you have just read, the orbits of Uranus and Neptune led to the discovery of Pluto. But the mass of Pluto is far too small, and therefore its gravitational pull too weak, to account for the unexpected orbits of Uranus and Neptune. Astronomers suspected that something else must be pulling on these planets, tugging them slightly from their expected orbits around the sun. Is this mysterious "something" a tenth planet? Astronomers have been looking for such a planet, nicknamed Planet X, which would be a giant planet some 8 billion kilometers beyond the orbit of Pluto.

What if no giant planet is found out there? Is there something else that might be tugging at Uranus and Neptune? Other possibilities exist. Many stars have a dark companion star. The sun may be part of such a binary-star system. Its dark companion could even be a brown dwarf, a massive object far larger than any planet but too small to have become a star. The brown dwarf would exert an enormous gravitational pull on the outer planets and might be found more than 80 billion kilometers from the sun.

Some astronomers have proposed an even more "far-out" explanation. They suggest that a black hole some 160 billion kilometers in space may be the source of the unexpected changes in the orbits of Uranus and Neptune. The black hole, at least ten times the mass of our sun, would exert a tremendous gravitational pull on these planets. Even from such a long distance it could reach into the solar system and disturb the orbits of Uranus and Neptune.

## Comets

You have gone a long way on your tour of the solar system and are now far beyond the outermost reaches of Pluto's orbit. You are about to visit the Oort cloud, named for the Dutch astronomer Jan Oort.

The Oort cloud is a vast collection of ice, gas, and dust some 15 trillion kilometers from the sun. Every once in a while, the gravitational pull of a

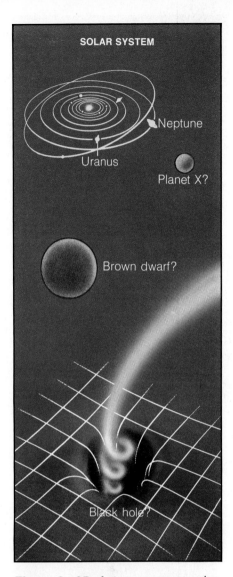

**Figure 2–35** *Astronomers wonder what unknown object may be tugging on the orbits of Uranus and Neptune. There are other possibilities—a tenth planet called Planet X, a brown dwarf, a black hole.*

**Figure 2–36** *Why does the tail of a comet always point away from the sun?*

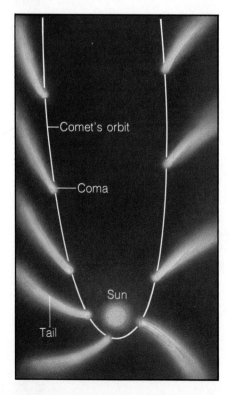

nearby star will tug a "dirty snowball" out of the Oort cloud and send it speeding toward the sun. For most of its trip toward the sun, this mountain-sized object, commonly called a **comet,** travels unnoticed. As it comes closer and closer to the sun, it grows warmer. Some of its ice, gas, and dust heat up enough to form a cloud around its core.

The core of a comet is called its nucleus. The cloud of dust and gas surrounding the nucleus is known as the coma. The nucleus and coma make up the head of the comet. During its approach to the sun, the head of the comet continues to grow warmer and to expand. In time, the head can expand to become as large as a few hundred thousand or even a million kilometers in diameter. Yet the head forms only a small part of the entire comet.

The sun produces a powerful solar wind made of high-energy particles. This solar wind blows the coma outward into a long tail that always streams away from the sun. The tail of an incoming comet streams out behind it. The tail of an outgoing comet streams in front of it. See Figure 2–36. In fact, the glowing tail was the basis for the term comet, which comes from the Greek word meaning long-haired. The tail of a comet often stretches out for millions of kilometers. A comet's tail is an astronomical wonder. It can sweep across a huge portion of the sky, yet it is so thin that distant stars can be seen shining through it.

Most of the 100,000 or so comets in the solar system orbit the sun over and over again. Many are long-period comets that have long, elliptical orbits, perhaps reaching out to the very edge of the solar system. Long-period comets may take thousands of years before they return to Earth's neighborhood again.

Short-period comets, however, return to the sun every few years. Perhaps the most famous short-period comet is Halley's comet, named for the English astronomer Edmund Halley. Halley's comet returns every 75 to 79 years. Although Halley did not discover the comet, he was the first to realize that the comets seen in 1456, 1531, 1607, and 1682 were really the same periodic comet. Halley predicted that the comet would return again in 1759, but he died without ever knowing whether his prediction

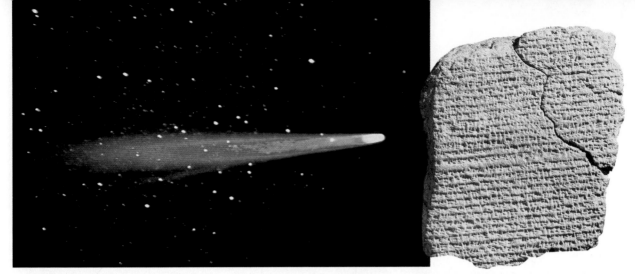

would actually come true. It did. The last time Halley's comet was seen was in 1986. It is due back again around 2062.

## Meteoroids, Meteors, and Meteorites

Earth is often "invaded" by objects from space. Most of these invaders are **meteoroids** (MEE-tee-uh-roids), chunks of metal or stone that orbit the sun. Scientists think that most meteoroids come from the asteroid belt or from comets that have broken up. Each day millions of meteoroids plunge through Earth's atmosphere. When the meteoroid rubs against the gases in the atmosphere, friction causes it to burn. The streak of light produced by a burning meteoroid is called a **meteor.** Meteors are also known as shooting stars.

Most meteors burn up in the atmosphere. A few, however, survive to strike Earth's surface. A meteor that strikes Earth's surface is called a **meteorite.** Meteorites vary in composition, but most contain iron, nickel, and stone.

While most meteorites are small, a few are quite large. The largest meteorite ever found is the Hoba West meteorite in South West Africa. It has a mass of more than 18,000 kilograms.

When a large meteorite crashes to Earth, it produces a crater. Some of the world's largest meteorite craters are found in Canada. In the United States, the most famous crater is the enormous Barringer Meteorite Crater, between Flagstaff and Winslow in Arizona.

**Figure 2–38** *This meteorite, discovered in Antarctica, is believed to have come from Mars.*

**Figure 2-39** *The Barringer crater in Arizona is 1.2 kilometers wide. What caused this huge crater?*

A meteorite found recently in Antarctica appears to have come from the moon. It is made of materials very similar to those brought back from the moon by astronauts. Even more exciting is an Antarctic meteorite that may have come from Mars. It appears similar in composition to the Martian soil tested by *Viking 1*. If so, it is the first known visitor from Earth's red neighbor in space.

## Life in the Solar System

In your mind, you have traveled across billions of kilometers as you explored the solar system. Yet nowhere in your travels have you come across living things. As far as scientists know, Earth is the only planet in our solar system that contains life. However, that does not mean that living things do not exist somewhere "out there."

For life as we know it to develop, certain conditions must be met. Two very important conditions are moderate temperatures and liquid water. And both must be present for billions of years for life to evolve. By chance, Earth has possessed these two conditions for most of its estimated 4.6-billion-year history. This is due partly to the fact that Earth happens to be in the very narrow "life zone" of its star, the sun.

If Earth had formed only 7.5 million kilometers closer to the sun, temperatures probably would have

**Figure 2–40** *The center illustration shows Earth as it is today. The illustration on the top shows what Earth might be like if its orbit were slightly farther from the sun. On the bottom is a view of Earth if its orbit were slightly closer to the sun.*

been too hot to support life. A location just 1.5 million kilometers more distant from the sun would have produced an Earth covered with frozen water. Again, life would not have developed. Yet these are simply probabilities. Even on Earth, living things have been found in the most improbable places.

## 2–3 Section Review

1. Briefly describe the major characteristics of the planets and other objects in the solar system.
2. What gives Mars its red color?
3. Which planet is thought to have lost its oceans due to the greenhouse effect?
4. Why is Neptune called the mathematician's planet?
5. Compare a meteoroid, meteor, and meteorite.
6. Why does a comet's tail always stream away from the sun?

**Connection—*Ecology***
7. During the process of photosynthesis, plants take in carbon dioxide from the atmosphere. Based on that information, why are scientists so concerned about the cutting down of huge patches of tropical rain forests on Earth?

# 2–4 Exploring the Solar System

Much of the information you have read in this chapter was provided by spacecraft sent to probe distant planets in our solar system. However, before a spacecraft can be sent to another planet, it has to be launched off the surface of the Earth. And to do

**Guide for Reading**

*Focus on these questions as you read.*

▶ *What is the principle behind the reaction engine used in rockets?*

▶ *What were the contributions of the various spacecraft sent by the United States to probe the solar system?*

that we need rockets. So we will begin our examination of the spacecraft we send to other planets by looking into the history of rocketry.

## Rocketry

Blow up a balloon and pinch the nozzle so that no air can escape. Hold the balloon at arm's length; then let it go. What happens?

When the balloon nozzle is released, air shoots out of it. At the same time, the balloon moves in a direction opposite to the movement of the escaping air. The released balloon is behaving like a rocket.

This example illustrates the idea of a **reaction engine.** Its movement is based on Sir Isaac Newton's third law of motion, which states that every action produces an equal and opposite reaction. The escaping rush of air out of the balloon nozzle causes the balloon to shoot off in the opposite direction. A reaction engine works in much the same way. **In a reaction engine, such as a rocket, the rearward blast of exploding gases causes the rocket to shoot forward.** The force of this forward movement is called thrust.

Long before Newton's time, the ancient Chinese, Greeks, and Romans made use of reaction engines. The Greeks and the Romans used steam to move toys. One toy consisted of a kettle on wheels with a basket holding glowing embers beneath it. Heat from the embers caused water in the kettle to boil. As the water boiled, steam hissed out of a horizontal nozzle on the kettle. In reaction to the escape of steam in one direction, the wheeled kettle rolled off in the opposite direction.

The first useful reaction engines were rockets developed by the Chinese around the year 1000. Their first known use was as weapons of war. These early Chinese rockets were long cylinders, probably sections of hollow bamboo, filled with gunpowder. One end of the cylinder was sealed, usually by a metal cap. The other end was open and had a fuse running through it into the gunpowder. When the gunpowder was ignited, burning gases shot out the open end of the cylinder. In reaction to this movement of gases, the cylinder shot off in the opposite direction.

By the end of the nineteenth century, some scientists began dreaming of using rockets to explore

**A**ctivity Bank

Action, Reaction, p.150

**Figure 2–41** *The action of the rocket's thrusters causes an opposite reaction and the rocket goes forward (top). Similar types of thrusters in this manned maneuvering unit (MMU) allow an astronaut to move in any direction in space (bottom).*

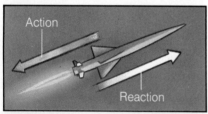

space. But would it be possible to build a rocket large enough and powerful enough to travel out of Earth's atmosphere? Late in the century a number of scientists studied this question. And at least one of them, a Russian named Konstantin E. Tsiolkovsky, considered it a definite possibility.

As a teenager, Tsiolkovsky had experimented with reaction engines. Using his lunch money to pay for materials, he had built a carriage powered by a reaction engine. But his experiment did not work. The engine could not develop enough thrust to move the carriage.

From his failure, Tsiolkovsky actually learned a lot about reaction engines. (Science is often like that.) He started to think about using such engines for space travel. Drawing on the work of Newton and other scientists and mathematicians, Tsiolkovsky worked out mathematical formulas for space flight. He even dreamed of creating human colonies in space. But before such colonies could be built, Tsiolkovsky knew that scientists would have to solve the enormous problems involved in building rockets powerful enough to escape the Earth's gravitational pull.

## Escape Velocity

In order for a rocket to escape Earth's gravitational pull, the rocket must achieve the proper velocity. This **escape velocity** depends on the mass of the planet and the distance of the rocket from the planet's center. The escape velocity from Earth is 11.2 kilometers per second, or 40,320 kilometers per hour. From the moon, it is just 2.3 kilometers per second. From mighty Jupiter, the escape velocity is 63.4 kilometers per second. Can you relate these differences in escape velocity to the mass of each planet?

The first step into space involves escape from the Earth. Tsiolkovsky predicted that through the use of a huge reaction engine, a vehicle would someday leave Earth's gravitational pull. But he also concluded that a rocket powered by gunpowder or some other solid fuel would not be able to accomplish this feat. Why not?

Solid fuels burn rapidly and explosively. The pushing force that results is used up within seconds.

**Figure 2–42** *This table shows the escape velocities for the nine planets in our solar system, the sun, and several other stars. Why is the escape velocity of Pluto so much lower than that of the other planets?*

| ESCAPE VELOCITIES | |
| --- | --- |
| Object | Escape Velocity (km/sec) |
| Mercury | 4.2 |
| Venus | 10.3 |
| Earth | 11.2 |
| Moon | 2.3 |
| Mars | 5.0 |
| Jupiter | 63.4 |
| Saturn | 39.4 |
| Uranus | 21.5 |
| Neptune | 24.2 |
| Pluto | 0.3* |
| Sun | 616 |
| Sirius B | 3400* |
| Neutron star | 200,000* |
| * Estimated | |

Although the force provides an enormous early thrust, it cannot maintain that thrust. As the rocket soars upward, the pull of Earth's gravity would tend to slow its climb and eventually bring it back to Earth.

In order for a rocket to build up enough speed to overcome the Earth's downward pull, the rocket must have a fuel that continues to burn and provide thrust through the lower levels of the atmosphere. Although Tsiolkovsky proposed this idea, he never built such a rocket. But in the 1920s, the American scientist Dr. Robert H. Goddard did. In 1926, Goddard combined gasoline with liquid oxygen and burned this mixture, launching a small rocket. The rocket did not go very far or very fast, but it did prove the point that liquid fuels could be used to provide continuous thrust.

Goddard built bigger and bigger rockets. And he drew up plans for multistage rockets. As each stage in such a rocket used up its fuel, the empty fuel container would drop off. Then the next stage would ignite, and its empty fuel container would drop off. In this way, a vehicle could be pushed through the atmosphere and out of Earth's grip. Today's rockets work in much the same way as Goddard's early rocket did. Now, however, the fuel is liquid hydrogen and liquid oxygen. Using such rockets, scientists are able to send spacecraft from Earth to the other planets in our solar system.

## Deep-Space Probes

The first spacecraft to travel beyond the solar system, in June 1983, was *Pioneer 10*. It was intended, along with *Pioneer 11*, to explore the outer planets of

**Figure 2–43** *You can see how the last stage of this rocket—the nose cone—is released into space.*

the solar system. In February 1990, seventeen years after it was launched, *Pioneer 11,* too, flew beyond the solar system.

**Pioneers, Vanguards, Explorers, Mariners, Rangers, Vikings, Surveyors, and Voyagers have been the workhorses of the effort of the United States to explore the solar system.** And their record has been impressive. The first successful probe of Venus was made by *Mariner 2* in 1962. The spacecraft approached to within 35,000 kilometers of the cloud-wrapped planet. *Mariner 2* quickly discovered that Venus, unlike Earth, does not have a magnetic field. And, as you have read earlier, Venus continues to be studied by spacecraft such as the *Magellan* probe.

The first successful probe of Mars was made by *Mariner 4,* in 1965. Going to within 10,000 kilometers of the red planet, the spacecraft sent back twenty-one photographs and other data. Two later probes, *Mariner 7* in 1969 and *Mariner 9* in 1971, sent back thousands of photographs of the Martian surface and the first detailed pictures of Mars's two moons.

The achievements of *Mariners 7* and *9* paved the way for the successful landings of *Vikings 1* and *2* on Mars in 1975, the first time spacecraft ever landed on another planet.

*Mariner 10* was the only spacecraft to fly by Mercury, innermost planet of the solar system. During three passes in 1974, *Mariner 10* mapped ancient volcanoes, valleys, mountains, and plains on the tiny planet.

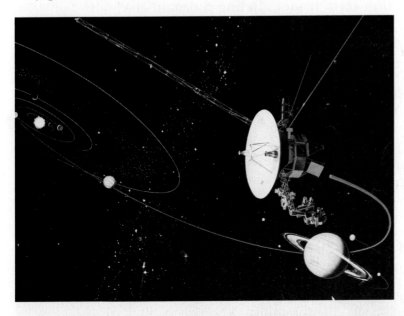

**Figure 2–44** *Here you see an artist's idea of how Voyager appeared as it passed by Saturn and made its way toward the very edge of our solar system and beyond.*

# ACTIVITY WRITING

## A Different Viewpoint

We view the solar system from the center of our world, which is Earth. Imagine how the solar system might look to intelligent creatures on another planet. In particular, how might those creatures view Earth? Use your ideas to write several pages that would be included in an alien's textbook on the solar system. (Of course, the book might be called *Neptune Science*—not *Earth Science*.)

Outward-bound *Pioneer 10* took a look at the giant planet Jupiter in December 1973 and sent back more than three hundred photographs. It also provided data on Jupiter's stormy atmosphere and its many moons. These findings were confirmed by the photographs sent back by *Pioneer 11* a short time later.

Six years later, two larger spacecraft, *Voyager 2* and *Voyager 1*, flew by Jupiter and sent back data that revealed surprises about the giant planet. Faint rings of particles and many new moons were discovered. The Pioneer and Voyager spacecraft examined Saturn and its ring system. Voyager photographs showed that what were considered to be a few broad rings are actually thousands of thin ringlets.

The missions of the Voyager spacecraft were far from over with the exploration of Saturn. Continuing on in the late 1980s and early 1990s, the Voyager spacecraft passed by and photographed Uranus and then Neptune, the twin giants of the outer solar system. Data from both spacecraft provided evidence regarding the atmosphere, core, and moons of these distant planets. Many scientists consider the Voyager spacecraft to be the most successful effort in the entire United States space program.

*Pioneer 10* and *Pioneer 11* are now beyond the solar system. By the end of this century, the Voyagers will follow into outer space. In case it should ever be found by people from another world, *Pioneer 10* contains a plaque with a message from the people of Earth. Who can say, but one day the most important discovery Pioneer may make is an advanced civilization on a planet circling a distant star!

## 2–4 Section Review

1. Identify at least one discovery made by each of the following spacecraft: Mariner, Pioneer, Viking, and Voyager.
2. How does the concept of action/reaction relate to rocketry?

**Connection—*You and Your World***
3. Discuss some of the ways sending spacecraft to distant planets has improved the quality of life for people on Earth.

## The First Magellan Probe

You began this chapter by reading about the exploration of Venus by the *Magellan* probe. The *Magellan* probe did not get its name by accident. It was named for one of the greatest explorers in *history*—Ferdinand Magellan.

Born in Portugal in the late fifteenth century, Magellan became convinced that a ship could sail around the world. In those days some people still did not believe the Earth was round, and they were convinced that any ship that sailed far enough would sail right off the edge of the Earth.

Before his famous trip to circle the world, Magellan studied astronomy and navigation for several years. Then in 1519, he commanded a fleet that set sail from Spain. The fleet traveled across the stormy Atlantic and reached the coast of Brazil. It then sailed down the coast of South America and wintered near what is now called Argentina. A mutiny broke out among Magellan's crew and several crew members were executed.

In 1520, the fleet set sail again and soon discovered a passage beneath South America that led them to a vast ocean, which Magellan called the Pacific Ocean, meaning peaceful ocean. To this day, the passage is called the Strait of Magellan.

Sailing across the Pacific was a great hardship for Magellan and his crew. Many suffered from a disease called *scurvy* caused by a lack of Vitamin C. Finally, the fleet reached the island of Guam. It was on Guam that Magellan suffered a fatal wound and died in 1521.

Magellan's crew continued the trip and eventually circled the planet by ship. And although Magellan did not live to see his dream completed, he is the first European to provide evidence that the Earth is a sphere.

Many historians consider Magellan's trip the greatest navigational feat in history. So it is not surprising that a twentieth-century spacecraft would be named after this courageous and daring explorer.

# Laboratory Investigation

## Constructing a Balloon Rocket

### Problem

How can a balloon rocket be used to illustrate Newton's third law of motion?

### Materials *(per group)*

drinking straw
scissors
9-m length of string
balloon
masking tape
meterstick

### Procedure 🔧

1. Cut the drinking straw in half. Pull the string through one of the halves.

2. Blow up the balloon and hold the end so that the air does not escape.

3. Have someone tape the drinking straw with the string pulled through it to the side of the balloon as shown in the diagram. Do not let go of the balloon.

4. Have two students pull the string tight between them.

5. Move the balloon to one end of the string. Release the balloon and observe its flight toward the other end of the string.

6. Record the flight number and distance the balloon traveled in a data table.

7. Repeat the flight of the balloon four more times. Record each flight number and length in your data table.

### Observations

1. What was the longest flight of your balloon rocket? The shortest flight?

2. What was the average distance reached by your balloon?

### Analysis and Conclusions

1. Using Newton's third law of motion, explain what caused the movement of the balloon.

2. Compare your balloon rocket to the way a real rocket works.

3. Suppose your classmates obtained different results for the distances their balloons traveled. What variables may have caused the differences?

4. **On Your Own** As you have read, rockets require a certain thrust to escape Earth's gravitational pull. How might you increase the thrust of your balloon rocket? Try it and see if you are correct.

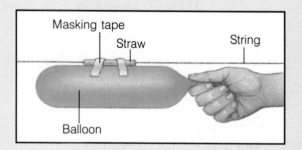

Masking tape    Straw    String
Balloon

# Study Guide

## Summarizing Key Concepts

### 2–1 The Solar System Evolves

▲ According to the nebular theory, the solar system formed from a huge cloud of gas and dust called a nebula.

### 2–2 Motions of the Planets

▲ A planet's period of revolution is the time it takes that planet to make one complete revolution around the sun, or a year on that planet.

▲ A planet's period of rotation is the time it takes that planet to make one complete rotation on its axis, or a day on that planet.

▲ The combined effects of inertia and gravity keep planets orbiting the sun in elliptical orbits.

### 2–3 A Trip Through the Solar System

▲ Mercury is a crater-covered world with high temperatures on its daylight side and low temperatures on its nighttime side.

▲ Venus is a cloud-covered world with high temperatures. The greenhouse effect is the main cause of its high temperatures.

▲ Mars is coated with iron oxide, or rust, which gives the planet its reddish color.

▲ The atmosphere of Jupiter is primarily hydrogen and helium.

▲ Jupiter's core is probably a rocky solid surrounded by a layer of liquid metallic hydrogen.

▲ Jupiter has sixteen moons.

▲ Saturn is similar in appearance and composition to Jupiter.

▲ Saturn's spectacular rings are made mostly of water ice.

▲ Uranus and Neptune are cloud-covered worlds. The atmosphere of both planets is primarily hydrogen, helium, and methane.

▲ The axis of Uranus is tilted at an angle of almost 90°. Uranus has at least fifteen moons and nine rings. Neptune has five known rings and eight known moons.

▲ Pluto is a moon-sized, ice-covered world with a large moon, Charon.

▲ The trail of hot gases from a burning meteoroid is called a meteor. If part of a meteoroid strikes the Earth, it is called a meteorite.

▲ As a comet approaches the sun, some of the ice, dust, and gas heat up and form a cloud around the nucleus.

### 2–4 Exploring the Solar System

▲ Much of the information about our solar system has been provided by spacecraft .

---

## Reviewing Key Terms

*Define each term in a complete sentence.*

### 2–1 The Solar System Evolves

solar system
nebular theory

### 2–2 Motions of the Planets

orbit
period of revolution
period of rotation

### 2–3 A Trip Through the Solar System

retrograde rotation
greenhouse effect
asteroid belt
magnetosphere
comet
meteoroid
meteor
meteorite

### 2–4 Exploring the Solar System

reaction engine
escape velocity

# Chapter Review

## Content Review

### Multiple Choice

*Choose the letter of the answer that best completes each statement.*

1. The outer gas giants do not include
   a. Jupiter.          c. Neptune.
   b. Pluto.           d. Saturn.
2. The observation that planets move in elliptical orbits was first made by
   a. Copernicus.     c. Kepler.
   b. Ptolemy.        d. Newton.
3. The time it takes a planet to make one complete trip around the sun is called its
   a. period of rotation.
   b. day.
   c. period of revolution.
   d. axis.
4. A planet with retrograde rotation is
   a. Venus.          c. Pluto.
   b. Jupiter.         d. Earth.
5. The Oort cloud is the home of
   a. asteroids.       c. meteorites.
   b. comets.         d. Pluto.
6. Rocklike objects in the region of space between the orbits of Mars and Jupiter are called
   a. comets.         c. asteroids.
   b. protoplanets.    d. meteorites.
7. The reddish color of Mars is due to
   a. carbon dioxide.  c. iron oxide.
   b. oxygen.         d. methane.
8. The planet that appears to be tipped on its side is
   a. Saturn.         c. Neptune.
   b. Uranus.         d. Venus.

### True or False

*If the statement is true, write "true." If it is false, change the underlined word or words to make the statement true.*

1. Due to the tremendous heat of the sun, the <u>outer</u> planets were unable to retain their lightweight gases.
2. The time it takes for a planet to travel once on its axis is called its <u>period of revolution</u>.
3. Newton recognized that planets do not sail off into space due to <u>gravity</u>.
4. The tail of a comet always streams <u>away</u> from the sun.
5. A <u>meteor</u> is a streak of light produced when a small object shoots through the atmosphere.
6. The <u>nebular theory</u> accounts for the formation of the solar system.
7. Newton's third law of motion states that for every action there is an <u>equal</u> and <u>opposite</u> reaction.

### Concept Mapping

*Complete the following concept map for Section 2–1. Refer to pages M6–M7 to construct a concept map for the entire chapter.*

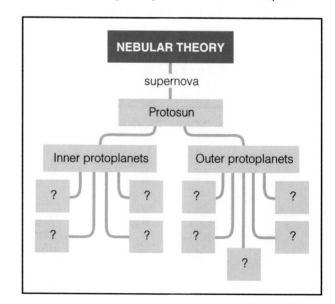

# Concept Mastery

*Discuss each of the following in a brief paragraph.*

1. Why do astronomers consider Jupiter a planet that was almost a star?
2. Describe the two types of planetary motion.
3. What factors led to the great differences between the rocky inner planets and the gaseous outer planets?
4. Describe the evolution of the solar system according to the nebular theory.
5. How has our understanding of the solar system been increased through mathematics?
6. Discuss the greenhouse effect in relation to Venus.

# Critical Thinking and Problem Solving

*Use the skills you have developed in this chapter to discuss each of the following.*

1. **Making comparisons** Compare the theories of Copernicus and Ptolemy.
2. **Relating cause and effect** Mercury is closer to the sun than Venus is. Yet temperatures on Venus are higher than those on Mercury. Explain why.

3. **Making predictions** Predict what the outer planets would be like if the sun were three times as large as it is.
4. **Making graphs** Using the chart on pages 68–69, draw a graph that plots the high and low temperatures on each planet. What conclusions can you draw from your graph?
5. **Making comparisons** Compare a meteoroid, meteor, and meteorite.
6. **Expressing an opinion** Sending spacecraft to probe the planets of our solar system costs many billions of dollars. Should the United States continue to spend money on space research, or could the money be better spent to improve conditions on Earth? What's your opinion?
7. **Using the writing process** Write a short story called "A Trip Around Planet Earth" in which you describe your home planet to an alien from another star system. Assume the alien has never been to Earth.

# *Earth* and *Its Moon*

## Guide for Reading

*After you read the following sections, you will be able to*

**3–1 The Earth in Space**

■ Relate Earth's rotation and revolution to day and night and to the seasons.

**3–2 The Earth's Moon**

■ Describe the characteristics of the moon.

■ Discuss several theories for the origin of the moon.

**3–3 The Earth, the Moon, and the Sun**

■ Identify the interactions among the Earth, the moon, and the sun.

**3–4 The Space Age**

■ Describe the functions of various types of artificial satellites.

■ Discuss some of the uses of space technology on Earth.

Hundreds of meters below the tiny spacecraft loomed a dusty plain strewn with boulders and craters—the strangest landscape humans had ever seen. But the two astronauts in the spidery craft had no time for sightseeing. Their job was to find a safe landing site—and fast, for they were running out of fuel. After traveling almost 400,000 kilometers from the Earth, they now had about 90 seconds to find a place to land on the moon.

"Down two and a half . . . forward, forward . . . good." Now they were just 12 meters above the plain called the Sea of Tranquility.

"Down two and a half . . . kicking up some dust." Nine meters to go!

"Four forward . . . drifting to the right a little." Finally, a red light flashed on the control panel.

"Contact light! Houston, Tranquility Base here. The *Eagle* has landed."

The date was July 20, 1969. For the first time in history, humans had left the Earth to explore its nearest neighbor in space, the moon. In the pages that follow, you will learn about the Earth's place in the solar system and about the relationship between the Earth and its moon.

## Journal *Activity*

***You and Your World***   Go outdoors on a clear night and look at the moon. What features can you see? In your journal, describe the appearance of the moon and include a sketch of what you see.

◄ *Astronaut Edwin E. Aldrin, Jr., walks on the surface of the moon. Notice the reflection of the Lunar Module* Eagle *in his faceplate.*

A *Foucault Pendulum Model*

In 1851, the French physicist Jean Foucault proved that the Earth rotates on its axis.

**1.** Tie a small weight such as an eraser to a piece of string. Tie the opposite end of the string to the arm of a ring stand.

**2.** Hang the pendulum over a turntable. The center of the turntable represents the North Pole. With a pen, make a reference mark on one side of the turntable.

**3.** Set the pendulum swinging and slowly turn the turntable. You will see that the direction of swing appears to change relative to the mark on the turntable.

If a Foucault pendulum is swinging above the North Pole, how long will it take for its direction of swing to appear to make one complete rotation? Explain.

# 3–1 The Earth in Space

Early observers saw the sun and the moon rise in the east and set in the west, just as we do today. Unlike modern observers, however, these people thought that the Earth stood still while the sun and the moon moved around it. Today we know that sunrise and sunset, as well as moonrise and moonset, are actually caused by the movements of the Earth in space.

Earth is the third planet from the sun in the solar system. Like all the other planets, Earth rotates on its axis as it travels around the sun. The Earth's axis is an imaginary line from the North Pole through the center of the Earth to the South Pole. And like the other planets, Earth revolves around the sun in an elliptical, or oval, orbit. **These two movements of the Earth—rotation and revolution—affect both day and night and the seasons on Earth.** Let's examine the effects of the Earth's rotation and revolution more closely.

## Day and Night

At the equator, the Earth rotates at a speed of about 1600 kilometers per hour. (This is about 400 kilometers per hour faster than the speed of sound in air.) It takes the Earth about 24 hours to rotate once on its axis. The amount of time the Earth takes

**Figure 3–1** *This dramatic photograph taken by the Apollo 11 astronauts shows the Earth rising above the moon's horizon.*

7:30 A.M.

10:30 A.M.

NOON

to complete one rotation is called a day. So a day on the Earth is about 24 hours long.

As the Earth rotates, part of it faces the sun and is bathed in sunlight. The rest of it faces away from the sun and is in darkness. As the Earth continues to rotate, the part that faced the sun soon turns away from the sun. And the part that was in darkness comes into sunlight. So the rotation of the Earth causes day and night once every 24 hours. Figure 3–2 shows a 12-hour sequence from sunrise to sunset.

If you could look down on the Earth from above the North Pole, you would see that the Earth rotates in a counterclockwise direction—that is, from west to east. The sun appears to come up, or rise, in the east, as the Earth turns toward it. The sun appears to go down, or set, in the west, as the Earth turns away from it. So a person standing on the rotating Earth sees the sun appear in the east at dawn, move across the sky, and disappear in the west at dusk.

You have probably noticed that throughout the year the length of day and night changes. This happens because the Earth's axis is not straight up and down. The Earth's axis is tilted at an angle of 23½°. If the Earth's axis were straight up and down, all parts of the Earth would have 12 hours of daylight and 12 hours of darkness every day of the year.

Because the Earth's axis is slightly tilted, when the North Pole is leaning toward the sun, the South Pole is leaning away from the sun. And when the South Pole is leaning toward the sun, the North Pole is leaning away from the sun. As a result, the

3:30 P.M.

7:30 P.M.

**Figure 3–2** This five-photo sunrise-to-sunset sequence was taken by a satellite orbiting the Earth above South America. What causes day and night on Earth?

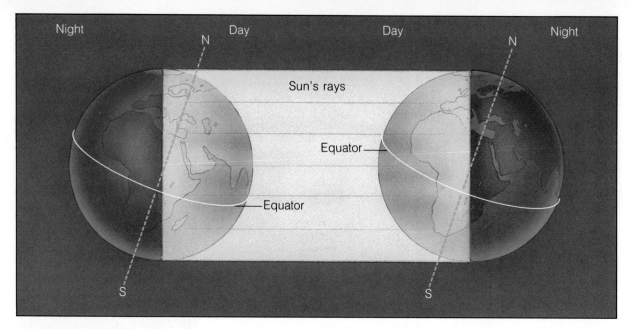

Night　　　　N　Day　　　　Day　　　　N　Night

Sun's rays

Equator

Equator

S　　　　　　　　S

**Figure 3–3** *Because of the tilt of the Earth's axis, the length of day and night is not constant. At the beginning of summer in the Northern Hemisphere (left), the North Pole is always in daylight and the South Pole is always dark. What happens at the beginning of winter (right)?*

# ACTIVITY

## DISCOVERING

*Temperature, Daylight, and the Seasons*

**1.** In your journal, keep a log of the high and low temperatures for each day in the school year.

**2.** Record the time at which the sun rises and sets each day.

**3.** Calculate the length of daylight, in hours and minutes, for each day.

■ What relationship do you see between the length of daylight and the temperature? Explain.

number of daylight hours in the Northern and Southern Hemispheres is not constant. (Recall that the Earth is divided into two halves, or hemispheres, by the equator.) The hemisphere that leans toward the sun has long days and short nights. The hemisphere that leans away from the sun has short days and long nights. Today, is the Northern Hemisphere leaning toward or away from the sun? How do you know?

## A Year on Earth

At the same time that the Earth is rotating on its axis, it is also revolving in its orbit around the sun. The Earth takes about 365.26 days to complete one revolution, or one entire trip, around the sun. The time the Earth takes to complete one revolution around the sun is called a year. So there are 365.26 days in one Earth year. How many times does the Earth rotate in one year?

You do not have to look at a calendar to know that there are only 365 days in a calendar year. But the Earth rotates 365.26 times in the time it takes to make one complete revolution around the sun. So about one fourth (0.26) of a day is left off the calendar each year. To make up for this missing time, an extra day is added to the calendar every four years. This extra day is added to the month of February, which then has 29 days instead of its usual 28. What is a year with an extra day called?

# PROBLEM Solving

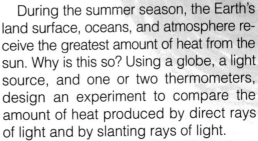

## What Causes Summer?

When the Northern Hemisphere is tilted toward the sun, that part of the Earth receives more direct rays of sunlight than the Southern Hemisphere does. The Southern Hemisphere, which is tilted away from the sun, receives slanting rays of sunlight. As a result, it is summer in the Northern Hemisphere and winter in the Southern Hemisphere.

During the summer season, the Earth's land surface, oceans, and atmosphere receive the greatest amount of heat from the sun. Why is this so? Using a globe, a light source, and one or two thermometers, design an experiment to compare the amount of heat produced by direct rays of light and by slanting rays of light.

**Finding cause and effect** Think about a typical sunny day. Do the sun's rays feel hotter at noon when the sun is directly overhead, or in the late afternoon when the sun is low in the sky? Explain.

## Seasons on Earth

Most people live in a part of the Earth that has four distinct seasons: winter, spring, summer, and autumn. If you could spend a year on each of the other eight planets in the solar system, you would find that five of them (Mars, Saturn, Uranus, Neptune, and possibly Pluto) share this characteristic with the Earth. The other three planets (Mercury, Venus, and Jupiter) either have no seasons at all or have seasons that vary so slightly they are not noticeable. Why do some planets have seasons and others do not?

If you study all the planets that have seasons, you will find that they all have one characteristic in common. They are all tilted on their axes. This is not true of the planets that do not have seasons. So you might conclude that the different seasons on Earth are caused by the tilt of the Earth's axis. And you would be correct.

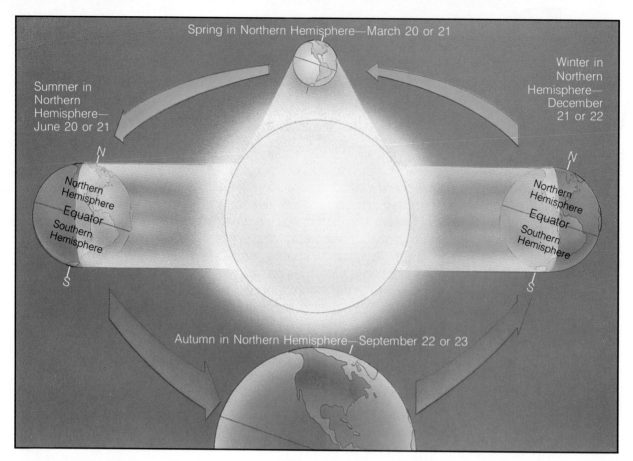

Spring in Northern Hemisphere—March 20 or 21

Summer in Northern Hemisphere— June 20 or 21

Winter in Northern Hemisphere— December 21 or 22

N

Northern Hemisphere

Equator

Southern Hemisphere

S

N

Northern Hemisphere

Equator

Southern Hemisphere

S

Autumn in Northern Hemisphere—September 22 or 23

**Figure 3–4** *When the North Pole is tilted toward the sun, the Northern Hemisphere receives more sunlight. It is summer. When the North Pole is tilted away from the sun, the Northern Hemisphere receives less sunlight. It is winter. Is the same true for the South Pole and the Southern Hemisphere?*

ACTIVITY

### CALCULATING

*The Earth on the Move*

The Earth moves at a speed of about 30 km/sec as it orbits the sun. What distance, in kilometers, does the Earth travel in a minute? An hour? A day? A year?

As the Earth revolves around the sun, the axis is tilted away from the sun for part of the year and toward the sun for part of the year. This is shown in Figure 3–4. When the Northern Hemisphere is tilted toward the sun, that half of the Earth has summer. At the same time, the Southern Hemisphere is tilted away from the sun and has winter. How is the Earth's axis tilted when the Southern Hemisphere has summer?

It is interesting to note that summer and winter are not affected by the Earth's distance from the sun. In fact, when the Northern Hemisphere is experiencing summer, the Earth is actually farthest away from the sun in its elliptical orbit. The same is true for summer in the Southern Hemisphere.

The hemisphere of the Earth that is tilted toward the sun receives more direct rays of sunlight and also has longer days than the hemisphere that is tilted away from the sun. The combination of more direct sunlight and longer days causes the Earth's surface and atmosphere to receive more heat from the sun. The result is the summer season.

Figure 3–5 *These photographs were taken on the same day in the Northern and Southern hemispheres. It is winter for the bison in Yellowstone National Park, Wyoming. But it is summer for the kangaroos in Australia. On this day, is the North Pole tilted toward or away from the sun?*

Summer begins in the Northern Hemisphere on June 20 or 21. This is the day when the North Pole is tilted a full 23½° toward the sun. The Northern Hemisphere has its longest day at this time, while the Southern Hemisphere has its shortest day. The longest day of the year is known as the **summer solstice** (SAHL-stihs). The word solstice comes from two Latin words meaning sun and stop. It refers to the time when the sun seems to stop moving higher in the sky each day. The sun reaches its highest point in the sky on the summer solstice.

After the summer solstice, the sun seems to move lower and lower in the sky until December 21 or 22, when the **winter solstice** occurs. At this time, the North Pole is tilted a full 23½° away from the sun. The shortest day of the year in the Northern Hemisphere and the longest day of the year in the Southern Hemisphere occur on the winter solstice.

Twice a year, in spring and in autumn, neither the North Pole nor the South Pole is tilted toward the sun. These times are known as equinoxes (EE-kwuh-naks-uhz). The word equinox comes from Latin and means equal night. At the equinoxes, day and night are of equal length all over the world. In the Northern Hemisphere, spring begins on the **vernal equinox,** March 20 or 21. Autumn begins on the **autumnal equinox,** September 22 or 23. What season begins in the Southern Hemisphere when spring begins in the Northern Hemisphere?

Figure 3–6 *Stonehenge, on Salisbury Plain in England, was built 4000 to 6000 years ago. Its massive stones were aligned to point to the rising or setting positions of the sun at the summer and winter solstices.*

# A Magnet in Space

Figure 3–7 *The Earth is enveloped in a huge magnetic field called the magnetosphere (bottom), whose lines of force produce the same pattern as those of a small bar magnet (top). The solar wind from the sun blows the magnetosphere far into space in the shape of a long tail.*

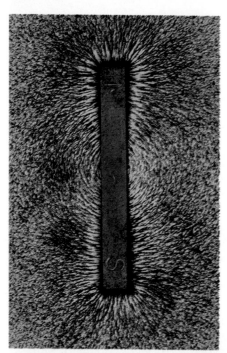

In Figure 3–7 you can see the pattern that forms when iron filings are sprinkled on a plate of glass covering a bar magnet. The pattern of iron filings reveals the invisible lines of force that connect the two poles, or ends, of the magnet. These invisible lines of force are called a magnetic field.

The Earth, much like a magnet, is surrounded by similar lines of force. In fact, you might think of the Earth as having a giant bar magnet passing through it from pole to pole. The Earth's magnetism forms a magnetic field around the Earth similar to the magnetic field around a bar magnet. Where does the Earth's magnetism come from? Although scientists are not sure, they think the Earth's magnetism is probably the result of the movement of materials in the Earth's inner core. The Earth's inner core is made mainly of the metals iron and nickel.

You can see in Figure 3–7 that the Earth's magnetic poles are not in the same place as the geographic North and South poles. The geographic poles are at the opposite ends of the Earth's tilted axis. But the magnetic poles, like the poles of a bar magnet, are at the ends of the lines of force that form the Earth's magnetic field.

The Earth's magnetic field is called the **magnetosphere.** The magnetosphere begins at an

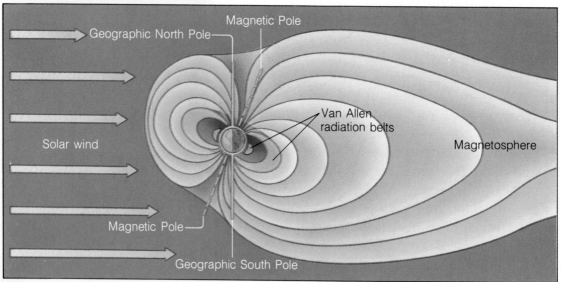

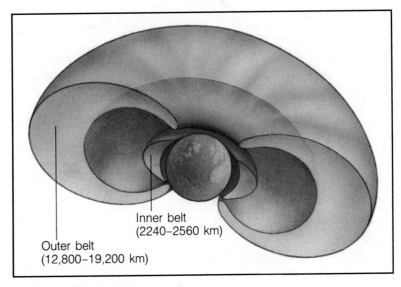

Inner belt
(2240–2560 km)

Outer belt
(12,800–19,200 km)

**Figure 3–8** *Particles in the solar wind are trapped to form the doughnut-shaped Van Allen radiation belts. What is it that traps the particles from the solar wind?*

altitude of about 1000 kilometers. It extends to an altitude of 64,000 kilometers on the side of the Earth facing toward the sun. But on the side of the Earth facing away from the sun, the magnetosphere extends in a tail millions of kilometers long! This long tail is caused by a stream of charged particles, called the solar wind, blowing from the sun. The solar wind constantly reshapes the magnetosphere as the Earth rotates on its axis. The magnetosphere is shown in Figure 3–7.

Two doughnut-shaped regions of charged particles are formed as the magnetosphere traps some of the particles in the solar wind. These regions are called the **Van Allen radiation belts.** They were named after Dr. James Van Allen, the scientist who first identified them. The outer Van Allen belt contains mostly electrons, or negatively charged particles. The inner Van Allen belt contains mainly protons, or positively charged particles.

Charged particles trapped by the Van Allen radiation belts may travel along magnetic lines of force to the Earth's poles. Here they collide with particles in the Earth's upper atmosphere. These collisions cause the atmospheric particles to give off visible light, producing an **aurora** (aw-RAW-ruh). The aurora may appear as bands or curtains of shimmering colored

**Figure 3–9** *The eerie green glow of the aurora australis, or southern lights, was photographed from the Space Shuttle* Discovery.

lights. Near the North Pole, the aurora is called the aurora borealis, or northern lights. Near the South Pole, the aurora is called the aurora australis, or southern lights.

## 3–1 Section Review

1. What causes day and night on the Earth?
2. How does the tilt of the Earth's axis, combined with the Earth's revolution, cause the occurrence of the seasons?
3. What do the words solstice and equinox mean? How are they related to the position of the Earth's axis?
4. What is the magnetosphere? What are the Van Allen radiation belts?
5. What causes an aurora?

**Connection—*You and Your World***
6. Describe the season you are experiencing today in terms of the Earth's position in space relative to the sun.

## Solar Wind Blows Out the Lights

Storms on the sun send forth huge bursts of charged particles. When these charged particles reach Earth, they can move the magnetic field that surrounds the Earth. When a magnetic field moves in relation to a conductor, an electric current results. In this case, the conductor is the Earth itself.

Usually, electric charges in the Earth equalize themselves by flowing back and forth through soil or rocks. Trouble results, however, when the electric current encounters dense igneous rocks. *Electricity* cannot flow easily through these rocks. As a result, the electric current seeks an easier path—and electric power lines provide the perfect alternative.

Because the equipment that manufactures electric power cannot handle the electric current produced by a solar storm, power systems can be seriously damaged or destroyed. This is what happened in March 1989 when a solar storm caused a power blackout in Quebec, Canada, and ruined two huge transformers at a nuclear power plant in southern New Jersey.

Due to the destructive potential of solar storms, several electric-power companies have suggested that a satellite "watchdog" be placed in orbit around the Earth. Such a satellite would be able to predict a solar storm in advance because charged particles in the solar wind travel much more slowly than electromagnetic radiation from the sun does. Sensors in the satellite would pick up signals from the electromagnetic radiation about an hour before the burst of charged particles actually took place. This advance warning would give power companies at least a chance to reduce power loads on certain transformers and possibly prevent sensitive equipment from being knocked out.

# 3–2 The Earth's Moon

"That's one small step for a man, one giant leap for mankind." With those words, astronaut Neil Armstrong became the first human to set foot on the moon. After landing, Armstrong and Edwin (Buzz) Aldrin left the Lunar Module *Eagle* to explore the moon. (The word lunar comes from the Latin word for moon.) While they walked on the surface, their fellow astronaut Michael Collins remained in orbit around the moon in the Apollo Command Module *Columbia*.

### Guide for Reading

*Focus on this question as you read.*

▶ *What are the major characteristics of the moon?*

**Figure 3–10** *Landing people on the moon was a complex mission. Here you see the Lunar Module approaching the moon (bottom). The Apollo Command Module (top), which remained in lunar orbit during the mission, was photographed from the Lunar Module shortly after it separated from the Command Module. After landing, the astronauts walked on the moon (right).*

Although Armstrong and Aldrin were the first astronauts to step onto the moon's dusty surface, they would not be the last. By the time the Apollo moon-landing project ended in 1972, twelve American astronauts had explored the moon. Scientists learned more about the moon from the Apollo missions than they had learned in the previous thousands of years. Here is some of what they now know about the moon.

## The Moon's Characteristics

The moon measures 3476 kilometers in diameter, or about one fourth the diameter of the Earth. It is much less dense than the Earth. The gravity of the moon is also less than that of the Earth. The moon's gravity is only one sixth that of the Earth. So objects weigh less on the moon than they do on Earth. To find out how much you would weigh on the moon, divide your weight by six. What would your weight be on the moon?

Today scientists know that the average distance to the moon is 384,403 kilometers. How do they know this? Among the instruments left on the moon by the Apollo astronauts was a small mirror. Scientists bounced a beam of laser light from the Earth off the mirror. Then they measured the amount of time it took the beam to bounce back to the Earth. Using

A CTIVITY

**READING**

*A Voyage to the Moon*

For a fictional account of the first exploration of the moon, read *From the Earth to the Moon* by the French novelist Jules Verne (1828–1905).

Figure 3–11 *The Apollo 14 astronauts left a mirror—the Laser Ranging Retro Reflector—on the surface of the moon. How did this mirror help scientists determine the distance to the moon?*

the known speed of light, they could then calculate the distance to the moon accurately.

Astronauts also left instruments on the moon to measure moonquakes. These instruments have measured as many as 3000 moonquakes per year. From these data, scientists now know that the moon's outer layer, or crust, is about 60 kilometers thick. Below the crust, there appears to be another layer of denser rock that is about 800 kilometers thick. Beneath this layer is a central core that may be made of melted iron.

Apollo astronauts also brought back samples of moon rocks. The oldest moon rocks are about 4.6 billion years old, which is about the same age as the Earth. So it seems likely that the moon and the Earth formed at about the same time.

The moon rocks brought back to Earth by the astronauts show no traces of water. So scientists believe that there never was any water on the moon. And the moon has no atmosphere. Without an atmosphere, there is no weather on the moon.

The temperature range on the moon is extreme. Noonday temperatures may rise above 100°C. During the long lunar nights, and in the shadows cast by crater walls, surface temperatures may drop to –175°C. **In short, the moon is a dry, airless, and barren world.**

## Features of the Moon

In 1609, Galileo Galilei became the first person to look at the moon through a telescope. He saw light areas and dark areas on the surface. The light areas he saw are mountain ranges soaring thousands of meters into the black sky. They are called **highlands.** Some of the highlands on the moon reach 8 kilometers above the surrounding plains. The broad, smooth lowland plains are the

**Figure 3–12** *The chart lists some important facts about the moon. How much greater is the moon's circumference than its diameter?*

## FACTS ABOUT THE MOON

**Average distance from Earth**
384,403 kilometers

**Diameter**
About 3476 kilometers (about $1/4$ Earth's diameter)

**Circumference**
About 10,927 kilometers

**Surface area**
About 37,943,000 square kilometers

**Rotation period**
27 days, 7 hours, 43 minutes

**Revolution period around Earth**
29 days, 12 hours, 44 minutes

**Length of day and night**
About 14 Earth-days each

**Surface gravity**
About $1/6$ Earth's gravity

**Mass**
$1/81$ Earth's mass

**Volume**
$1/50$ Earth's volume

**Figure 3–13** *Astronaut James Irwin explores the moon in a Lunar Rover. The surface of the moon has changed very little since Galileo studied it in 1609. What evidence of their presence did the Apollo astronauts leave on the moon?*

dark areas Galileo saw through his telescope. He called them **maria** (MAHR-ee-uh). *Maria* (singular, *mare*) is the Latin word for seas. Although the name maria seems to indicate that there is water on the moon, scientists now know that the moon has no surface water. Why do you think Galileo called the plains maria?

Among the most striking features of the moon are its many craters. Craters ranging in size from microscopic to hundreds of kilometers across cover the moon's surface. One of the largest craters on the moon is called Copernicus. (Many craters are named for famous scientists. Copernicus was a Polish astronomer who first stated the theory that Earth and the other planets revolve around the sun.) The crater Copernicus is approximately 91 kilometers in diameter. Imagine a crater about the same distance across as the distance between Houston, Texas, and Galveston, Texas. Most of the moon's craters are in the highlands. Few craters are located in the maria.

Scientists think that most craters were formed by the continuous bombardment of meteorites. These meteorites blasted out craters when they hit the moon. A few of the craters, however, seem to be the result of volcanic activity. In fact, the maria are filled with hardened rock that may have flowed into the plains as hot lava billions of years ago. And lava comes from active volcanoes.

**Figure 3–14** *Here you see a moon rock brought back to Earth by the Apollo astronauts. What have scientists learned about the moon by studying these rocks?*

**Figure 3–15** *Meteorites have been crashing into the moon for billions of years and have carved out many craters. Here you see the large crater Tsiolkovsky on the far side of the moon (right).*

Other evidence that the moon once had active volcanoes may be the long valleys, or **rilles,** that crisscross much of its surface. Hadley Rille is one such valley. It is about 113 kilometers long. Because there was no running water to carve the rilles, they might have been cut into the moon's surface by rivers of flowing lava. Another possible explanation is that the rilles may be cracks caused by moonquakes. Or the rilles may have opened up billions of years ago when the moon's hot surface cooled and shrank. Whatever caused the rilles, scientists agree that today's cold and inactive moon was once hot and active.

## Movements of the Moon

As the Earth revolves around the sun, the moon revolves around the Earth in an elliptical orbit. At **perigee** (PEHR-uh-jee), the point of the moon's orbit closest to the Earth, it is about 350,000 kilometers from the Earth. At **apogee** (AP-uh-jee), the point of the moon's orbit farthest from the Earth, it is about 400,000 kilometers away.

Like the sun, the moon seems to move west across the sky. This apparent movement is caused by the Earth's rotation. When viewed against a background of stars, however, the actual movement of the moon eastward can be observed. You can

**Figure 3–16** *This photograph of the far side of the moon clearly shows some of the thousands of craters scattered across the moon's surface.*

prove this by observing the moon when it is at the western edge of a cluster of stars. If you observe carefully for several hours, the moon's eastward movement can be seen as it passes in front of each star, one after the other.

Recall that it takes the Earth about 24 hours to rotate once on its axis. The moon takes much longer. The moon rotates once on its axis every 27.3 days. This is the same amount of time it takes the moon to revolve once around the Earth. Thus the moon's period of rotation is the same as its period of revolution. This means that a day on the moon is just as long as a year on the moon! As a result, the same side of the moon always faces toward the Earth. For many years, scientists could observe only the side of the moon that faces the Earth. Then the Lunar Orbiter space probe photographed the far side of the moon for the first time. The astronauts of the Apollo 8 mission, which circled the moon, were the first humans to see the far side directly. The entire surface of the moon has now been photographed. These photographs show a bleak, lifeless landscape of boulders, craters, plains, and valleys.

## Origin of the Moon

One of the most interesting questions scientists have asked about the moon is: Where did the moon come from? There are several theories concerning the moon's origin. According to one theory, the moon may have been formed millions or billions of kilometers away from the Earth and later "captured" by the Earth's gravity. Most scientists, however, do not think that is what happened. Another theory is that the moon formed from the same swirling cloud of gas and dust from which the sun, Earth, and other planets were formed. Indeed, the composition of the moon is similar enough to the composition of the Earth to indicate that they could have formed from the same material.

Probably the most likely explanation for the origin of the moon is that the moon was "born" when a giant asteroid the size of the planet Mars struck the early Earth, tearing a chunk of material from the planet. According to this theory, the Pacific Ocean may be the hole left when the moon was torn from the Earth. This explanation, based on evidence from moon rocks, would explain why the moon is so similar to Earth and yet has no water. Any water in the material that was blasted off the Earth would have been vaporized (turned into a gas) when the material was blown into orbit. When the material came together again to form the moon, there would have been no water left.

Scientists are still not sure how the moon was formed. But clues gathered by astronauts and robot space probes are helping scientists to solve the mystery of the moon's origin.

# 3–2 Section Review

1. What are the main characteristics of the moon?
2. How might the moon's craters and rilles have formed?
3. Why is a day on the moon the same length as a year on the moon?
4. Describe three possible theories to explain the origin of the moon.

**Critical Thinking—*Making Inferences***
5. Why is the distance between the Earth and the moon usually given as an average?

# 3–3 The Earth, the Moon, and the Sun

We can think about the motions of the Earth and the motions of the moon separately. Usually, however, scientists consider the motions of the Earth–moon system as a whole. As the Earth moves in its yearly revolution around the sun, the moon moves in its monthly revolution around the Earth. At the same time, both the Earth and the moon rotate on their axes. Gravitational attraction keeps the moon in orbit around the Earth and the Earth in orbit around the sun. **The relative motions of the Earth, the moon, and the sun result in the changing appearance of the moon as seen from the Earth and the occasional blocking of the sun's light.**

## Phases of the Moon

The moon revolves around the Earth. The revolution of the moon causes the moon to appear to change shape in the sky. The different shapes are called phases of the moon. The phases of the moon are shown in Figure 3–17. The moon goes through all its phases every 29.5 days.

Why does the moon go through phases? The moon does not shine by its own light. Rather, the moon reflects sunlight toward the Earth. The phase of the moon you see depends on where the moon is in relation to the sun and the Earth. Refer to Figure 3–17 as you read the description that follows.

When the moon comes between the sun and the Earth, the side of the moon facing the Earth is in darkness. The moon is not visible in the sky. This phase is called the new moon. Sometimes the new moon is faintly visible because it is bathed in earthshine. Earthshine is sunlight reflected off the Earth onto the moon.

As the moon continues to move in its orbit around the Earth, more of the lighted side of the moon becomes visible. First a slim, curved slice called a crescent appears. This is the waxing crescent phase. The moon is said to be waxing when the lighted area appears to grow larger. When the lighted area appears to grow smaller, the moon is said to be waning.

*Observing the Moon*

The best time to observe the moon is on the second or third day after the first-quarter phase. At this time, the moon is in a good position in the evening sky. Many surface features are clearly visible because the moon is not reflecting full light toward the Earth. Details are easier to see.

**1.** Using binoculars or a small telescope and a labeled photograph of the moon, locate some of the most prominent surface features on the moon.

**2.** Draw a sketch of the moon and label the features you were able to identify.

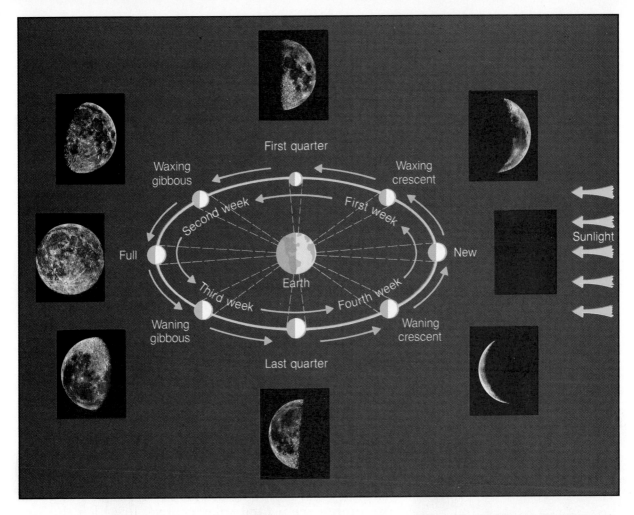

Figure 3–17 *The eight phases of the moon you see from Earth depend on where the moon is in relation to the sun and the Earth. How many days does it take for the moon to pass through all eight phases?*

About a week after the new moon, the moon has traveled a quarter of the way around the Earth. At this time, half the moon appears lighted. This phase is the first-quarter phase. As the days pass, more of the lighted area can be seen during the waxing gibbous phase.

About two weeks after the new moon, the entire lighted side of the moon is visible in the sky. This phase is called the full moon. The Earth is then between the moon and the sun. The moon takes another two weeks to pass through the waning-gibbous, last-quarter, and waning-crescent phases, and back to the new moon. The phases of the moon then start all over again. What phase of the moon was visible to you last night?

## Eclipses

More than 2000 years ago, a great war was about to begin. The two armies faced each other across a vast plain. Suddenly, the sky turned dark. The sun seemed to be swallowed up. Both armies thought the sun's disappearance was a sign that they should not fight. Although the sun soon reappeared, the frightened soldiers went home without fighting. There was no battle that day. The soldiers never knew that what they had witnessed was not a sign but one of Earth's most dramatic natural events—an eclipse.

As the moon revolves around the Earth and the Earth and the moon together revolve around the sun, they occasionally block out some of the sun's light. The long, cone-shaped shadows that result extend thousands of kilometers into space. Sometimes the moon moves into the Earth's shadow. At other times the moon casts its shadow onto the Earth. Each event results in an eclipse.

There are two types of eclipses. The two types of eclipses are named depending on which body—the sun or the moon—is eclipsed, or blocked. A **solar eclipse** occurs when the new moon comes directly between the sun and the Earth. As the Earth moves into the moon's shadow, sunlight is blocked from reaching the Earth. Any shadow has two parts. The small, inner shadow is called the **umbra** (UHM-bruh). The larger, outer shadow is called the **penumbra** (pih-NUHM-bruh). Only people directly in the path of the umbra see a total solar eclipse, in which the sun is completely blocked out. People in the penumbra see a partial solar eclipse, in which only part of the sun is blocked out.

As you have read, a new moon occurs once a month. But as you probably know from experience, this is not true for a solar eclipse. Why is there not a solar eclipse once every month? The answer is that the orbit of the moon is slightly tilted in relation to the orbit of the Earth. A solar eclipse takes place only when the new moon is directly between the Earth and the sun, which happens only rarely.

To anyone who has ever seen a total solar eclipse, the experience is awe-inspiring. When the sky begins to darken, birds are fooled into thinking it is evening, and so they stop singing. Dogs begin to howl. The air temperature drops sharply. For a few minutes, day becomes night. If you are ever fortunate enough to view a solar eclipse, you must remember one very important rule. Never look directly at the sun. Your eyes may be burned by

**Figure 3–18** *The magnificent pearly light of the sun's corona is visible during a total solar eclipse.*

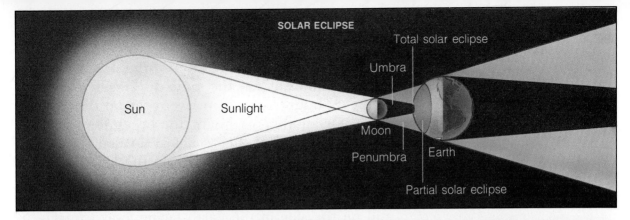

**SOLAR ECLIPSE**

Total solar eclipse

Umbra

Sun    Sunlight

Moon

Penumbra    Earth

Partial solar eclipse

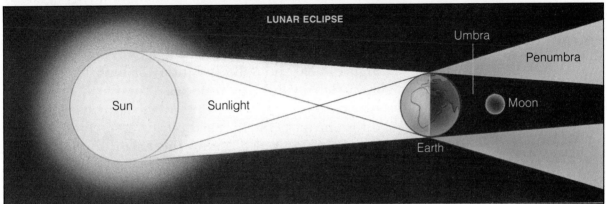

**LUNAR ECLIPSE**

Umbra

Penumbra

Sun    Sunlight

Moon

Earth

the sun's rays—even when they do not appear to be visible—and you may be blinded.

The second type of eclipse occurs when the Earth comes directly between the sun and the full moon. This event is called a **lunar eclipse.** A lunar eclipse takes place when the moon passes through the Earth's shadow. When the moon moves through the umbra, a total lunar eclipse occurs. When the moon moves through the penumbra, a partial lunar eclipse occurs. Earth's shadow falling on the full moon dims the moon's glow to a dark coppery color. This eerie reddish color results when sunlight reflected off the moon is bent as it passes through Earth's atmosphere.

## Tides

Because the moon is close to the Earth, there is a gravitational attraction between the Earth and the moon. As a result of the gravitational pull of the Earth on the moon, the side of the moon facing the Earth has a distinct bulge. But the moon also exerts a gravitational pull on the Earth. This pull results in the rise and fall of the oceans as the moon moves in its orbit around the Earth.

**Figure 3–19** *During a solar eclipse, the moon passes between the sun and the Earth. During a lunar eclipse, the Earth passes between the sun and the moon. When does a total eclipse of the moon occur?*

**A**ctivity Bank

What Causes High Tides?, p.151

If you have ever spent a day at the beach, you probably noticed that the level of the ocean at the shoreline did not stay the same during the day. For about six hours, the ocean level rises on the beach. Then, for another six hours, the ocean level falls. The rise and fall of the oceans are called **tides.**

As the moon's gravity pulls on the Earth, it causes the oceans to bulge. The oceans bulge in two places: on the side of the Earth facing the moon and on the side of the Earth facing away from the moon. Each of these bulges causes a high tide.

At the same time that the high tides occur, low tides occur between the two bulges. The diagram in Figure 3–20 shows the positions of high tides and low tides on the Earth. Because the Earth rotates on its axis every 24 hours, the moon's gravity pulls on different parts of the Earth at different times of the day. So at a given place on the Earth, there are two high tides and two low tides every 24 hours. But because the moon rises about 50 minutes later each day, the high and the low tides are also 50 minutes later each day.

Some high tides are higher than other high tides. For example, during the full moon and the new moon phases, the high tides are higher than at other times. These higher tides are called **spring tides.** Spring tides occur because the sun and the moon are in a direct line with the Earth. The increased

**Figure 3–20** *The pull of the moon's gravity causes tides. During a 12-hour period, Los Angeles moves from a high tide (left) to a low tide (center) to another high tide (right). At the same time, the moon moves in its orbit. As a result, the tides occur slightly later each day.*

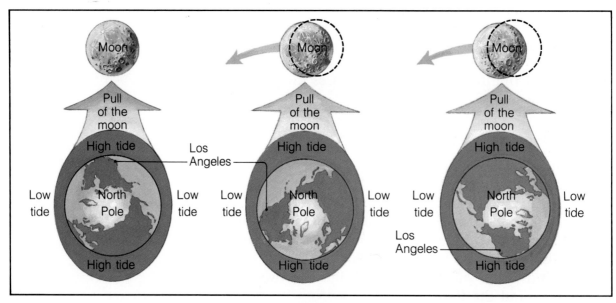

**Figure 3–21** *At high tide, boats float serenely at their moorings in the Bay of Fundy in Nova Scotia (left). A few hours later, at low tide, the boats are left sitting in the mud (right).*

effect of the sun's gravity on the Earth causes the ocean bulges to be larger than usual.

When the moon is at the first and last quarter phases, its gravitational pull on the oceans is partially canceled by the sun's gravitational pull. This results in high tides that are lower than usual. These lower high tides are called **neap tides.** What is the position of the sun and the moon with respect to each other during neap tides?

The varying distance between the Earth and the moon as the moon moves in its orbit also affects the tides. The closer the moon is to the Earth, the greater the pull of the moon's gravity on the Earth. If the moon is at perigee during a new moon or a full moon, extremely high tides and low tides will occur.

## 3–3 Section Review

1. How is the relative motion of the Earth, the moon, and the sun related to the phases of the moon? To eclipses?
2. When would you be able to see a lunar eclipse? A solar eclipse?
3. How does the moon affect tides on the Earth?

**Critical Thinking—*Relating Concepts***
4. Why does a lunar eclipse occur either two weeks before or two weeks after a total solar eclipse?

# 3–4 The Space Age

The first exciting step into space was taken on October 4, 1957. On that historic day, a Soviet rocket boosted *Sputnik 1,* the world's first artificial satellite, into Earth orbit. The Space Age had begun!

Since that day, thousands of artificial satellites have been placed in orbit around the Earth. Astronauts have gone to the moon and returned. People have lived and worked in Earth orbit for extended periods of time. Slowly, humans are taking their first tentative steps beyond the home planet and into the solar system. At the same time, we are learning more about our planet and finding ways to improve our lives.

## Artificial Satellites

Artificial satellites are satellites built by people. Like the moon, which is Earth's natural satellite, artificial satellites travel just fast enough so that they neither escape the Earth's gravity nor fall back to the Earth's surface. There are several different types of satellites orbiting the Earth. **Communications satellites, weather satellites, navigation satellites, and scientific satellites are among the artificial satellites that orbit the Earth today.** Each of these types of satellites has a specific function.

**Figure 3–22** *In spite of problems with its main mirror, astronomers hope the* Hubble Space Telescope *will reveal many secrets about the universe. This view is from inside the Space Shuttle as the telescope was being held by the robot arm before being released from the cargo bay.*

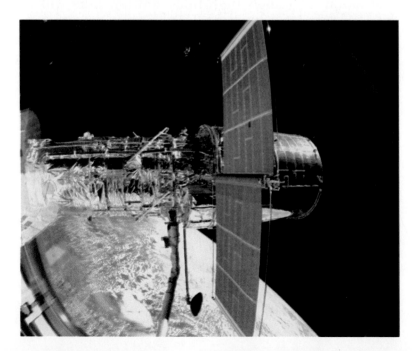

**COMMUNICATIONS SATELLITES** Many of the satellites orbiting the Earth are communications satellites. Communications satellites beam television programs, radio messages, telephone conversations, and other kinds of information all over the world. You can think of a communications satellite as a relay station in space. The satellite receives a signal from a transmitting station on Earth. The satellite then beams the signal to a receiving station somewhere else on Earth. In this way, information is quickly transmitted from one place to another, even on the other side of the world.

Communications satellites are often placed in **geosynchronous** (jee-oh-SIHNG-kruh-nuhs) **orbit.** A geosynchronous orbit is one in which the satellite revolves around the Earth at a rate equal to the Earth's rotation rate. As a result, the satellite stays in one place above a certain point on the Earth's surface. Three such satellites placed equal distances apart at an altitude of about 35,000 kilometers above the Earth can relay signals to any place on Earth.

**WEATHER SATELLITES** Artificial weather satellites have greatly improved our ability to track weather patterns and forecast weather conditions all over the world. By studying and charting weather patterns, scientists can predict the weather with greater accuracy than ever before. These predictions are particularly important in tracking dangerous storms such as hurricanes. Through the use of weather satellites, scientists today can better predict when and where a hurricane will strike. This information gives people in the path of a hurricane time to protect themselves and their property and often saves lives.

**NAVIGATION SATELLITES** Navigation satellites are another type of artificial satellite. They send precise, continuous signals to ships and airplanes. Using information from navigation satellites, sailors and pilots can determine their exact locations within seconds. This information is especially useful during storms when other kinds of navigation equipment may not provide accurate information. Someday it may even be possible for cars and trucks to use navigation satellites to pinpoint their locations.

**SCIENTIFIC SATELLITES** Many different types of scientific satellites now orbit the Earth. Long before

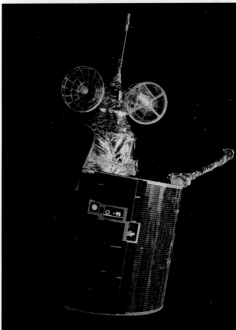

**Figure 3–23** *A communications satellite floats above Africa after being released into orbit by the Space Shuttle* Columbia *(top). The GOES satellite makes it possible for meteorologists to track the paths of severe storms (bottom). What type of satellite is GOES?*

Figure 3–24 *Advance warning of a hurricane provided by weather satellites helps to prevent property damage and may even save lives. Manhattan is visible in the center of this satellite map of New York City and the surrounding area.*

*Sputnik,* scientists looked forward to the day when they could observe the Earth and the universe from orbiting satellites. They thought scientific satellites would help them to solve old mysteries and to make new discoveries about the universe. And they were correct!

In 1958, the first satellite launched by the United States, *Explorer 1,* discovered the Van Allen radiation belts around the Earth. Since that time, other scientific satellites have added greatly to our knowledge of the universe. One satellite in particular—the Infrared Astronomy Satellite, or *IRAS*—has solved many mysteries of the universe. *IRAS* has found evidence of planetary systems forming around distant stars, as well as evidence of massive black holes at the center of the Milky Way and other galaxies. (A black hole is a collapsed star that is so dense that nothing—not even light—can escape its grip.)

Other scientific satellites focus their attention on Earth. In 1991, the Upper Atmosphere Research Satellite was sent into orbit to study the Earth's protective ozone layer. This satellite was the first in a series of environmental science satellites launched as part of Mission to Planet Earth. Mission to Planet Earth is a long-term research program that will use scientific satellites to study the Earth's environment.

## Laboratories in Space

In 1973, the United States launched Skylab into orbit. Skylab was a space station designed to allow astronauts to perform experiments in space. Astronauts could dock their spacecraft with Skylab and enter its laboratory. Later the astronauts could re-enter their spacecraft and return to Earth with their data. Visiting crews aboard Skylab made detailed studies of the sun, conducted several health-related experiments, and learned to work in space. Skylab was the United States' first laboratory in space.

Today Spacelab has taken the place of Skylab. Spacelab is a laboratory that is designed to be carried into orbit by the Space Shuttle. The laboratory can be fitted with different types of scientific equipment, depending on the types of experiments being performed.

In 1986, the Soviet Union launched a space station called *Mir*. (The Russian word *mir* means peace.) Space station *Mir* was designed as a series of modules that could be added to the original basic module. Today, *Mir* consists of four permanent modules. Eventually, two more modules may be added on. Cosmonauts, who often remain on the space station for months, perform a variety of scientific experiments aboard *Mir*.

*Mir* is now the largest and most complex space station ever to orbit the Earth. The United States, however, is planning to build its own space station, possibly as a joint venture with Russia, Japan, and several other nations. The space station will be built in orbit from parts carried into space by the Space Shuttle. Someday the space station may serve as a base for return trips to the moon and for the exploration of Mars.

**Figure 3–25** *The* Mir *space station orbits 320 kilometers above the Earth (bottom right). Here you see an artist's idea of what a United States space station might look like (bottom left). Astronaut Mae Jemison performs experiments in the science module of Spacelab J inside the cargo bay of the Space Shuttle* Endeavour *(top).*

**Figure 3–26** *Multilayer insulation was developed to protect experiments in the Space Shuttle's open cargo bay. It is now used to make cold weather clothing. The Newtsuit, an experimental diving suit made possible by space technology, will be used as a model for twenty-first century space suits. The bicyclist is wearing a miniature insulin pump first developed for NASA.*

## Space Technology Spinoffs

**Although most of the major discoveries of the space program have been made far from the Earth, many aspects of space technology have practical applications.** Because these applications have been "spun off" the space program, they are called space technology spinoffs. They owe their existence to the exploration of space. Thousands of spinoffs—from heart pacemakers to lightweight tennis rackets—have resulted from applications of space technology.

In 1967, NASA scientists and engineers searched for a fabric to use in spacesuits. The fabric would have to be strong enough to withstand the extreme temperature variations in space and yet flexible enough to fashion into a spacesuit. It was not long before such a fabric was invented. Astronauts walking in space and on the moon found themselves dressed properly—and safely—for conditions in space. Soon after, the same fabric was used to make roofs for a department store in California, an entertainment center in Florida, and a football stadium in Michigan.

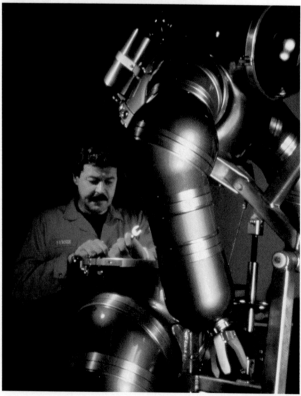

An astronaut exposed to direct sunlight in space runs the risk of overheating. To reduce the danger of overheating, space scientists developed various devices to be fitted into spacesuits. One of these devices was a gel packet that draws excess heat away from the body. These gel packets are now used by marathon runners to absorb excess heat from their foreheads, necks, and wrists.

One of the questions that puzzled scientists was how the human body would react to the new environment of space. To find the answers, scientists designed a series of automatic monitoring devices that would relay to Earth an astronaut's blood pressure, heart rate, and other vital statistics. Such devices are now used by paramedics when they answer emergency calls. These devices provide rapid and accurate information about a patient's condition. Such information can often mean the difference between life and death.

The message seems clear. Space technology—even if it is intended for use far beyond the frontiers of Earth—may have practical applications for billions of people who will never get farther from the Earth's surface than an energetic leap can take them.

## ACTIVITY WRITING

*Space Technology Facilities*

Space-technology facilities are located throughout the world. Some of those run by NASA in the United States are listed below. Choose one that sounds interesting to you and find out more about that facility. Write a brief report of your findings. Be sure to include where the facility is located, what technological development occurs there, and what the plans are for future research.

Goddard Space Flight Center
Jet Propulsion Laboratory
Kennedy Space Center
Langley Research Center
Lewis Research Center
Marshall Space Flight Center

## 3–4 Section Review

1. What are four kinds of artificial satellites? Describe the basic function of each.
2. How can space technology be beneficial to people? Why are applications of space technology called spinoffs?
3. Describe one practical application for each of the following: the fabric used in spacesuits; gel packets used to keep astronauts from overheating; automatic monitoring devices used to keep track of an astronaut's vital signs.

**Critical Thinking—*Applying Concepts***
4. Explain what you think is meant by the following statement: "The space program is a down-to-Earth success."

# Laboratory Investigation

## Observing the Apparent Motion of the Sun

### Problem

How can the sun's apparent motion in the sky be determined by observing changes in the length and direction of a shadow?

### Materials *(per student)*

wooden stick and base
piece of cardboard, 25 cm x 25 cm
compass
wide-tip felt pen
metric ruler

### Procedure 🔲

1. Place the stick attached to a base in the middle of a piece of cardboard. Trace the outline of the base on the cardboard so that you will be able to put it in the same position each time you make an observation of the sun.

2. Place the stick and the cardboard on flat ground in a sunny spot.

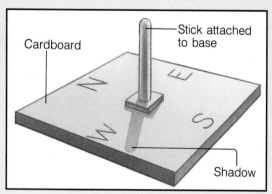

Cardboard

Stick attached to base

Shadow

3. Using the compass, locate north, south, east, and west. Write the appropriate directions near the edges of the cardboard.

4. With the felt pen, trace the shadow of the stick on the cardboard. Write the time of day along the line. Measure the length of the shadow. Determine in which direction the shadow is pointing. Determine the position of the sun in the sky. **CAUTION:** *Do not look directly at the sun!* Record your observations in a data table similar to the one shown here.

5. Repeat step 4 five more times throughout the day. Be sure to include morning, noon, and afternoon observations.

### Observations

1. In which direction does the sun appear to move across the sky?

2. In which direction does the shadow move?

3. At what time of day is the shadow the longest? The shortest?

### Analysis and Conclusions

1. Why does the length of the shadow change during the day?

2. What actually causes the sun's apparent motion across the sky?

3. **On Your Own** How is it possible to tell time using a sundial? Turn your shadow stick into a sundial by writing the correct time of day in the appropriate places on the cardboard.

| Time of Day | Shadow Length | Direction of Shadow | Location of Sun |
|---|---|---|---|
| | | | |
| | | | |
| | | | |
| | | | |

## Summarizing Key Concepts

### 3–1 The Earth in Space

▲ The rotation and revolution of the Earth affect both day and night and the seasons on Earth.

▲ The apparent motions of the sun and the moon in the sky are caused by the rotation of the Earth.

▲ The tilt of the Earth's axis, combined with its revolution, causes the seasons.

▲ The Earth is surrounded by a magnetic field called the magnetosphere.

### 3–2 The Earth's Moon

▲ The moon has neither water nor an atmosphere.

▲ The main features of the moon are highlands, maria, craters, and rilles.

▲ The moon revolves around the Earth in an elliptical orbit.

▲ There are three possible theories to explain how the moon was formed.

### 3–3 The Earth, the Moon, and the Sun

▲ The relative motions of the Earth, the moon, and the sun result in the phases of the moon and in eclipses.

▲ As the moon revolves around the Earth, its shape appears to change; the moon goes through all its phases in 29.5 days.

▲ There are two types of eclipses: solar eclipses and lunar eclipses.

▲ The gravitational pull of the moon on the Earth causes the tides.

### 3–4 The Space Age

▲ Artificial satellites include communications satellites, weather satellites, navigation satellites, and scientific satellites.

▲ Space stations, such as Skylab, Spacelab, *Mir,* and *Freedom,* serve as laboratories in space.

▲ Space technology spinoffs have many practical applications on Earth.

## Reviewing Key Terms

*Define each term in a complete sentence.*

### 3–1 The Earth in Space
summer solstice
winter solstice
vernal equinox
autumnal equinox
magnetosphere
Van Allen radiation belts
aurora

### 3–2 The Earth's Moon
highlands
maria
rille
perigee
apogee

### 3–3 The Earth, the Moon, and the Sun
solar eclipse
umbra
penumbra
lunar eclipse
tide
spring tide
neap tide

### 3–4 The Space Age
geosynchronous orbit

# Chapter Review

## Content Review

### Multiple Choice

*Choose the letter of the answer that best completes each statement.*

1. Smooth lowland areas on the moon are
   a. maria.
   b. rilles.
   c. highlands.
   d. craters.
2. The phase of the moon that follows the waning-crescent phase is called the
   a. full moon.
   b. new moon.
   c. waxing crescent.
   d. last quarter.
3. The sun reaches its highest point in the sky on the
   a. summer solstice.
   b. winter solstice.
   c. vernal equinox.
   d. autumnal equinox.
4. Which of the following are examples of space technology spinoffs?
   a. heart pacemakers
   b. heat-absorbing gel packets
   c. blood-pressure monitors
   d. all of these
5. The magnetic field around the Earth is called the
   a. solar wind.
   b. aurora australis.
   c. magnetosphere.
   d. aurora borealis.

6. The Earth's axis is tilted at an angle of
   a. 90°.
   b. 45°.
   c. 23½°.
   d. 30°.
7. Of the following, the one that is a natural satellite of the Earth is
   a. the moon.
   b. a communications satellite.
   c. a scientific satellite.
   d. a weather satellite.
8. The first satellite launched by the United States was
   a. *IRAS.*
   b. Skylab.
   c. *Explorer 1.*
   d. *Sputnik.*
9. Every four years, an extra day is added to the month of
   a. January.
   b. February.
   c. March.
   d. December.
10. Another name for the aurora borealis is the
    a. magnetosphere.
    b. Van Allen belts.
    c. northern lights.
    d. southern lights.

## True or False

*If the statement is true, write "true." If it is false, change the underlined word or words to make the statement true.*

1. The Earth rotates in a <u>clockwise</u> direction.
2. When the Northern Hemisphere is tilted toward the sun, it is <u>summer</u> in the Southern Hemisphere.
3. The Earth's magnetic poles <u>are</u> in the same place as the geographic poles.
4. The longest day of the year occurs on the <u>winter</u> solstice.
5. The moon goes through all its phases every <u>30</u> days.
6. The outer part of a shadow is called the <u>umbra</u>.
7. Exceptionally high tides that occur during a full-moon phase are called <u>neap</u> tides.

## Concept Mapping

*Complete the following concept map for Section 3–1. Refer to pages M6–M7 to construct a concept map for the entire chapter.*

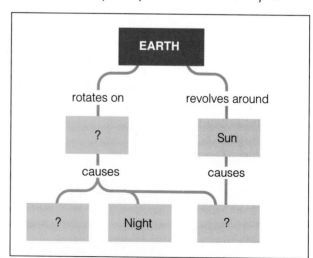

# Concept Mastery

*Discuss each of the following in a brief paragraph.*

1. Explain why one side of the moon always faces toward the Earth while the other side always faces away from the Earth.
2. Describe the functions of the various kinds of artificial satellites.
3. How is the direction of the Earth's rotation related to the apparent motion of the sun across the sky?
4. What is the difference between a solstice and an equinox?
5. Explain why there are tides on the Earth.
6. Why should you never look directly at the sun?
7. Why is an Earth year (the time it takes the Earth to complete one revolution around the sun) different from a calendar year? What is done to make up for this difference?
8. What is the difference between a solar eclipse and a lunar eclipse? When can you see a total solar eclipse? A total lunar eclipse?

# Critical Thinking and Problem Solving

*Use the skills you have developed in this chapter to answer each of the following.*

1. **Applying concepts** Lunar eclipses occur during a full moon. The moon goes through a full-moon phase every month. Yet lunar eclipses are fairly rare. Why is there not a lunar eclipse every month? *Hint:* You may want to use models to arrive at your answer.
2. **Making predictions** Describe how living conditions on the Earth might change if the Earth's axis were straight up and down instead of tilted.
3. **Interpreting diagrams** Identify each of the phases of the moon in the diagram.
4. **Identifying relationships** The moon seems to move westward across the sky. But when it is viewed against the background of stars, it appears to move eastward. Explain why this is so.
5. **Making comparisons** How is the Earth like a magnet?
6. **Using the writing process** Write a science-fiction story describing what you think life would be like on an orbiting space station or a moonbase in the twenty-first century.

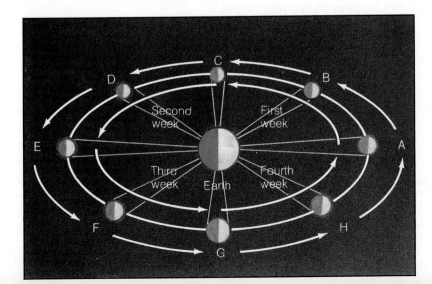

# IAN K. SHELTON DISCOVERS AN

The time: 170,000 years ago. Much of North America is covered by huge sheets of ice. Herds of woolly mammoths and other unusual creatures roam the land. From time to time, the ancestors of modern humans gaze up at the stars twinkling in the night sky. Although these human ancestors have no way of knowing, in a galaxy 170,000 light years away a giant star is exploding.

The time: February 24, 1987. Ian K. Shelton, a young Canadian scientist, prepares to spend another long night at the Cerro Tololo Inter-American Observatory in Chile, South America. Shelton assumes it will be another quiet night. But little does he know that light produced when that giant star exploded more than 170,000 years ago will finally reach the Earth this night!

Shelton has been studying photographs of a small galaxy called the Large Magellanic Cloud. By early morning, he is ready to call it a night. "I had decided," he recalls, "that enough was enough. It was time to go to bed." Yet before going to sleep, Shelton decided to develop one last photograph.

As he studied the last photograph, Shelton realized there was something most unusual in the picture. A bright spot could be seen. Photographs taken of the same area on previous nights had not shown this bright spot. "I was sure there was some flaw on the photograph," he recalled. But then he did something astronomers rarely do. He went outside and looked up at the area of the sky he had just photographed. And without the telescope, or even a pair of binoculars, Shelton saw the same bright spot in the Large Magellanic Cloud. He knew right away that this was something new and unusual.

Shelton could hardly believe what he was seeing. "For more than three hours," he explained later, "I tried several logical

# EXPLODING S*T*A*R

explanations. It took me a long time to actually accept that what I had just seen was a supernova."

Supernovas are the last stage in the life of certain giant stars. As the star dies it begins to contract. Then, in its last moments of life, the star explodes and sends matter and energy blasting through the universe.

During a supernova, a star reaches temperatures of billions of degrees Celsius. At those temperatures, atoms in the star fuse and new elements are produced. The light

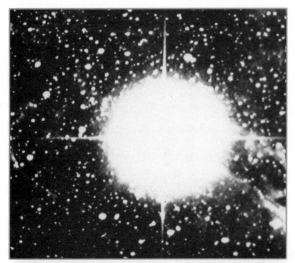

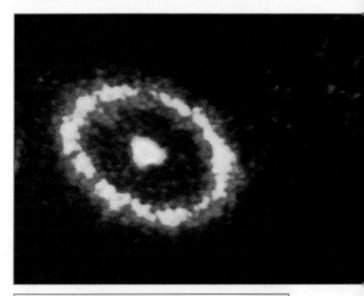

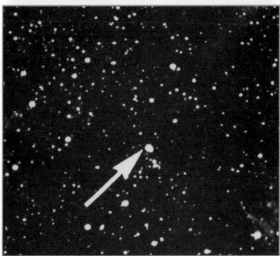

◄▲  On February 24, 1987, Ian K. Shelton observed the supernova pictured top left. The bottom left photograph, taken three years before, shows the star that became Supernova 1987A. The photograph above shows a view of Supernova 1987A taken by the Hubble Space Telescope in 1990. The green ring is gas released before the star exploded and the pink blob is debris from the explosion.

produced by a supernova is brighter than the light produced by a million normal stars. It was that bright light Shelton observed in 1987, after the light had traveled 170,000 years toward Earth.

Shelton immediately sent telegrams to astronomers all over the world. Observatories in other parts of the world soon confirmed Shelton's discovery. "It's like Christmas," remarked astronomer Stan Woosley from the University of California. This was the first supernova close to the Earth that modern astronomers had ever had a chance to study. A few weeks after the discovery, the new supernova was officially named Supernova 1987A.

Why, you might wonder, is the discovery of a new supernova so important? Many as-

tronomers believe that supernovas cause the birth of new stars. So, by studying supernovas, astronomers can learn a great deal about the life cycles of stars. Also, the elements a supernova produces shower nearby areas of space. In fact, most of the elements on the Earth probably formed some 6 billion years ago during a supernova. "The calcium in our bones, the iron in hemoglobin and the oxygen we all breathe came from explosions like this one," says astronomer Woosley.

For Ian K. Shelton, the discovery of an exploding star would change his life. Shelton knows he owes some of the credit for the discovery to modern technology. "We couldn't conduct modern astronomy without these wonderful instruments," he has said. "But without the romance, most of us would never have been attracted to this wonderful science in the first place. Just look at that beautiful supernova up there. Isn't that enough to make you glad you're alive?"

# LOOKING FOR LIFE BEYOND EARTH

## *Should the Search for ETs Go On?*

▲ **Although this extraterrestrial of movie fame is a fictional character, some scientists believe that intelligent life may exist somewhere else in the universe.**

O ur sun is only one of more than billions and billions of stars in the universe. Scientists believe that many of these stars have planets, more than half of which are probably larger than Earth. That makes our planet a fairly ordinary member of a very large group.

Yet we know our planet has a very special feature. It is the home of intelligent life. We are on it! Does that make Earth unique in all the universe? Are we alone, or could there be other intelligent beings among the stars?

According to Frank J. Tipler, an American physicist, "Earth is unique. We are alone. Extraterrestrial beings, living things not from Earth, do not exist." A number of biologists share Tipler's view. They say that a combination of complex conditions and circumstances—"rare accidents"—led to the evolution of intelligent life on Earth. In the opinion of these scientists, the odds are against the same events occurring anywhere else in the universe. These scientists argue further that, if we were not alone, we would have been visited or contacted by beings from another world by now. They suggest that the absence of contact with extraterrestrials, or ETs, is strong evidence that such beings do not exist.

Many scientists, including several prominent astronomers, disagree with these arguments. They think it is quite unlikely that Earth is the only planet in the vast universe on which intelligent life might have developed. Although we cannot be sure that intelligent beings exist elsewhere in the universe, they say, we certainly cannot be sure they do not exist. In the words of American astronomer Carl Sagan, "absence of evidence is not evidence of absence."

## THE SEARCH

Professor Woodruff T. Sullivan III of the University of Washington points out that the only way to silence skeptics and be sure that extraterrestrial intelligence exists is to find direct evidence of it. Over many years, scientists in several countries have taken part in a search for extraterrestrial intelligence, or SETI. These scientists have been listening for special radio signals from space.

There has been no shortage of signals. The universe is a very noisy place in which stars, planets, and many other objects give off a broad range of radio energy waves. However, none of these signals form patterns that would indicate that they carry a message from ETs.

What makes the search even more complicated is the fact that scientists are not sure which radio wavelengths to "eavesdrop" on, nor are they sure of where to aim their antennae. In other words, the scientists do not know what "station" to tune in. Moreover, they are not even sure they would recognize an intelligent signal if they did tune in the right "station." Would such a signal have to be a regular pattern of electromag-

▲ With its giant reflecting dish pointed toward the heavens, the Goldstone radio telescope searches for a message from an extraterrestrial civilization. If such a message is ever received, it would be the most monumental discovery in human history.

netic radiation, such as a series of "dots and dashes"? Or might it appear totally random to a human observer? As Professor Frank Drake of Cornell University says, the present search for ETs is like sifting for an "alien needle" in a "cosmic haystack."

The idea of using radio telescopes to search for signals from space was first proposed in 1959 by Massachusetts Institute of Technology professor Philip Morrison. Morrison realized that radio signals traveling at the

# ISSUES IN SCIENCE

▲ **SETI pioneer Philip Morrison, along with co-author Guiseppe Cocconi, was the first scientist to suggest searching for radio signals at the wavelength of hydrogen.**

speed of light—about 300,000 kilometers per second—were the only practical way for alien civilizations to communicate with Earth. The problem now was to narrow down the range of radio frequencies to search. Morrison quickly realized that hydrogen, which is the most abundant element in the universe, radiates energy at a frequency of 1420 megahertz. He decided that this was the "magic frequency" at which to search for signals from space. "I was convinced," Morrison explains, "that an advanced extraterrestrial civilization, one smart enough to communicate across the galaxies with radio technology, would certainly also have detected . . . the hydrogen signal."

## WHY SEARCH?

There are many different motives behind the search for ETs. Many people believe that it would be comforting to know that we are not alone in the universe. We would come to see the whole universe as, in some sense, our home. At the same time, we would recognize that we must live cooperatively and not as though we humans were in charge of the universe.

Some people also believe that highly developed alien civilizations could teach us a great deal about technology. Enlightened aliens might provide new energy sources, advanced medical cures, and even the ability to travel to the stars. Perhaps they could also help us in dealing with problems of communication among humans.

Even if we do not succeed in contacting ETs, the effort devoted to the search could still produce positive technological and social results. For example, it could lead to new developments in radio astronomy. In particular, it could increase our knowledge of the sources of the many different types of radio signals streaming in from space.

Much of the opposition to the search for extraterrestrials comes from people who either are convinced that intelligent life exists only on Earth or feel that extraterrestrial beings can never be contacted from Earth. These people strongly object to spending money on SETI programs.

However, as part of a National Aeronautics and Space Administration (NASA) program, three giant radio antennae normally used to communicate with spacecraft are being used to listen for radio signals from space. The antennae are located at Goldstone, California; Madrid, Spain; and Tidbinbilla, Australia. Connected to the telescopes will be a radio receiver that can scan at least 10 million radio channels at the same time!

The scientists involved in the SETI programs don't expect any easy discoveries. They foresee technical problems, increased skepticism, disappointments, and budget battles ahead. However, these setbacks and problems are not likely to curb the basic human curiosity that prompts the question, Are we alone? As long as some people think there is a possibility that we are not alone and there is a chance, no matter how remote, that we can contact another civilization, the search will probably go on.

# SCIENCE GAZETTE:

# VOYAGE TO THE
# RED PLANET
## ESTABLISHING THE FIRST COLONY ON MARS

**T**hirty-three days from Earth, the explorers aboard the *Martian Mayflower* are showing signs of space fatigue. Some are angry and hostile much of the time. Others are depressed and withdrawn. Almost everyone suffers occasional headaches and nausea. Weightlessness, crowded living conditions, and lack of privacy have made the space voyagers tense and weary.

The voyagers wonder silently whether they will be able to endure another month cooped up in the spacecraft. However, they know they have no choice. The *Martian Mayflower* has passed the point of no return. It is more than halfway to planet Mars.

## TO CREATE A DISTANT COLONY

Sixty days from Earth, the surface of Mars looms ahead. The mood of the space voyagers is changing. They are excited by the thought of stepping onto a new world and creating the first human settlement beyond Earth.

Mars is not a totally unknown world to the people of the *Martian Mayflower*, however. Some years earlier, three astronauts explored an area of the Martian surface. When they returned to Earth, the astronauts brought back samples of Martian rock and atmosphere. Careful study of these samples convinced scientists that people could settle on Mars.

In addition to the three human astronauts, hundreds of robot instruments, including

△ **At speeds topping 35,000 kilometers per hour, the spaceship finally approaches Mars. Feelings of boredom and weariness are forgotten, quickly replaced by a sense of awe.**

▲ Colonists search for ice in a huge Martian canyon more than ten times as long as the Grand Canyon on Earth.

some robot vehicles, have been sent to Mars to prepare for the *Martian Mayflower* expedition. Several of these robots were activated on landing. They are broadcasting essential information, such as reports on Martian weather and surface conditions, to the approaching space voyagers. Other robots wait silently on Mars. Instruments and supplies on these robots may mean the difference between life and death for the settlers during the long Martian year, which lasts 669 days.

## TOUCHDOWN

The landing site selected for the *Martian Mayflower* is on a plateau above a deep canyon. The plateau is relatively level and smooth with few large boulders. It seems to be ideal for the spaceship touchdown. Moreover, instruments aboard previous spacecraft have detected signs of underground ice, a possible source of drinking water.

The location of the canyon was also an important factor in picking the landing site. Fierce winds and dust storms frequently sweep across the surface of Mars. A deep canyon seemed the best place to shelter the expedition's living quarters and scientific

labs from the harsh Martian weather.

As the spaceship orbits Mars, robot instruments on the Martian surface are measuring wind velocity, temperature, and atmospheric pressure. This information is transmitted to the spaceship's master computer. The computer then determines the best path to the landing site and the safest speed for the descent to the surface.

At last, the calculations are complete. The spaceship angles into position. On computer command, maneuvering rockets fire to slow the spaceship and send it plunging into the thin Martian atmosphere. Close to the surface, the maneuvering rockets fire again to correct the spaceship's path of descent and position it over the landing site. Special parachutes open to ease the spaceship onto the Martian plateau. The landing is soft and smooth. The ship is undamaged and will be able to return the colonists to Earth when their work is finished.

## DIGGING IN

Shortly after touchdown, the voyagers give their protective spacesuits one last check. Finally, groups of explorers are ready to leave the ship and set foot on the dusty Martian surface. Each group has an appointed task.

The first group sets out to find a Mars rover vehicle. The position of the robot vehicle, which had landed a few months earlier, was pinpointed just before the spaceship landed. At that time, the robot vehicle seemed close to the landing site. But as the spacesuited voyagers struggle over the rocky Martian surface, the robot vehicle seems far away. The surrounding area reminds the voyagers of the dry valley in Antarctica where they trained. However, there are differences.

On Mars, dust particles give the atmosphere a rusty tinge. Occasionally the voyagers stumble into kneehigh dust drifts. Hours pass before the weary voyagers finally reach the robot vehicle.

Meanwhile, back near the landing site, several groups search for subsurface deposits of ice. At the spaceship, the remaining voyagers struggle to build a solar power station and temporary living quarters. Since gravity on Mars is one-third that on Earth, the voyagers find that they can lift and move relatively large loads. So within a short period of time, they are able to set up a power station, a life-support system, and living quarters around the spaceship.

Giant mirrors in orbit around Mars collect solar energy and beam it to the power station in the form of microwaves. In the station, the microwaves are converted into electricity, which is used in the life-support system to concentrate Martian atmospheric gases. Carbon dioxide is liquefied and removed. The result is a breathable mixture of nitrogen, argon, and oxygen with traces of water vapor.

If it is absolutely necessary, the colonists can survive by condensing this water vapor into liquid water. But it will not be necessary. Two of the groups searching for ice report success. Large deposits of subsurface frozen water and carbon dioxide have been located near the landing site. These deposits can be mined and processed to provide water, oxygen, hydrogen, and plant nutrients.

Late in the Martian day, the colonists get more good news. The first group of explorers rolls into camp aboard a functioning Mars rover.

## THE COLONY SUCCEEDS

As the days pass, the Mars rover is used to collect materials and instruments from other robot craft. From these materials, the voyagers construct permanent quarters. In time, the voyagers will create a large area where they can work, study, relax, and exercise without their cumbersome spacesuits.

As the colonists learn to use the various resources of Mars for their own benefit and that of future colonists, they also gain enormous knowledge of the Red Planet. But most of all, they know that humans can survive, and even thrive, far beyond Earth's protective atmosphere.

▽ **The first human settlement beyond the Earth provides voyagers the opportunities to work, study, relax, and exercise.**

# For Further Reading

If you have been intrigued by the concepts examined in this textbook, you may also be interested in the ways fellow thinkers—novelists, poets, essayists, as well as scientists—have imaginatively explored the same ideas.

## Chapter 1: Stars and Galaxies

Anker, Charlotte. *Last Night I Saw Andromeda*. New York: Henry Z. Walck.

Clarke, Arthur C. *2001: A Space Odyssey*. New York: New American Library.

Engdahl, Sylvia. *This Star Shall Abide*. New York: Atheneum.

Ride, Sally, with Susan Okie. *To Space and Back*. New York: Lothrop, Lee & Shepard.

## Chapter 2: The Solar System

Cameron, Eleanor. *The Wonderful Flight to the Mushroom Planet*. Boston: Little Brown and Co.

Gallant, Roy A. *The Constellations: How They Came to Be*. New York: Four Winds.

Harris, Alan, and Paul Weissman. *The Great Voyager Adventure: A Guided Tour Through the Solar System*. Englewood Cliffs, NJ: Julian Messner.

Jones, Diana Wynne. *Dogsbody*. New York: Greenwillow.

## Chapter 3: Earth and Its Moon

Del Rey, Lester. *Prisoners of Space*. Philadelphia: Westminster Press.

Heinlein, Robert A. *Rocket Ship Galileo*. New York: Charles Scribner's Sons.

Lawrence, Louise. *Moonwind*. New York: Harper & Row.

Verne, Jules. *Journey to the Center of the Earth*. New York: New American Library.

# Activity Bank

Welcome to the Activity Bank! This is an exciting and enjoyable part of your science textbook. By using the Activity Bank you will have the chance to make a variety of interesting and different observations about science. The best thing about the Activity Bank is that you and your classmates will become the detectives, and as with any investigation you will have to sort through information to find the truth. There will be many twists and turns along the way, some surprises and disappointments too. So always remember to keep an open mind, ask lots of questions, and have fun learning about science.

**Chapter 1    STARS AND GALAXIES**

| | |
|---|---|
| ALL THE COLORS OF THE RAINBOW | **146** |
| SWING YOUR PARTNER | **147** |
| HOW CAN YOU OBSERVE THE SUN SAFELY? | **148** |

**Chapter 2    THE SOLAR SYSTEM**

| | |
|---|---|
| RUSTY NAILS | **149** |
| ACTION, REACTION | **150** |

**Chapter 3    EARTH AND ITS MOON**

| | |
|---|---|
| WHAT CAUSES HIGH TIDES? | **151** |

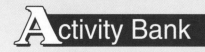
# ALL THE COLORS OF THE RAINBOW

Stars come in many different colors—from blue stars, to yellow stars such as the sun, all the way to red stars at the opposite end of the spectrum. The visible light emitted by stars is also made up of different colors. To study starlight, astronomers use a spectroscope. A spectroscope breaks up light into its characteristic colors. In this activity you will build a simple spectroscope.

## Materials

shoe box
scissors
cardboard
tape
diffraction grating
black construction paper
uncoated light bulb

## Procedure

1. Carefully cut two small, square holes in opposite ends of a shoe box.

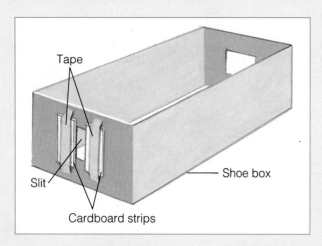

2. Tape two small pieces of cardboard on either side of one hole to make a narrow slit.

3. Tape a piece of diffraction grating over the other hole. **Note:** *Before you tape the diffraction grating in place, hold it up in front of a light. Turn the diffraction grating so that the light spreads out into a horizontal spectrum.*

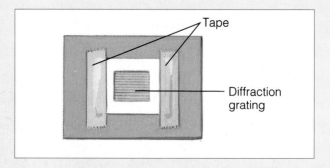

4. Cover the inside of the shoe box, except for the two holes, with black construction paper. Then tape the shoe box closed.

5. Hold your spectroscope so that the slit is parallel to the bright filament of an uncoated light bulb. Look at the light and describe what you see. **CAUTION:** *Do not point your spectroscope at the sun. Never look directly at the sun.*

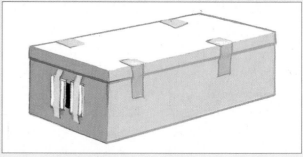

## Going Further

If a fluorescent light bulb or a neon light is available, look at it through your spectroscope and describe its spectrum.

# Activity Bank

# SWING YOUR PARTNER

Gravity is the force of attraction between all objects in the universe. The more mass an object has, the stronger its gravitational attraction. The Earth has the largest mass of any nearby object, so we are always aware of the Earth's gravity. On Earth, gravity keeps our feet firmly on the ground! Gravity also causes falling bodies to accelerate, or change their velocity, as they fall toward the Earth's surface. The acceleration caused by the Earth's gravity is equal to 1 *g*. In this activity you will measure the value of *g* in meters per second per second (m/sec$^2$).

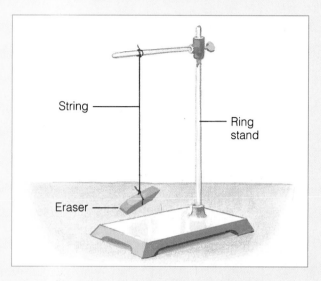

## Materials

string
metric ruler
eraser
ring stand
clock or watch with second hand

## Procedure

1. Tie an eraser to a piece of string about 50 cm long.

2. Make a pendulum by tying the free end of the string to the arm of a ring stand. Record the length of the string, in meters, in a data table similar to the one shown.

3. Pull the eraser to one side and release it. Count the number of complete swings the eraser makes in 60 sec. ·Record this number in your data table.

4. Use the following equation to find the period (*T*) of the pendulum: *T* = 60 sec/number of swings. Record the period, in seconds, in your data table.

5. Repeat steps 3 and 4 three more times. Find the average period of the pendulum.

6. Calculate the gravitational acceleration *g* using the following formula:

$$g = 4\pi^2L/T^2$$

In this formula, $\pi = 3.14$, *L* is the length of the pendulum in meters, and *T* is the average period of the pendulum in seconds. What value did you find for *g*?

## DATA TABLE

| Trial | Length (m) | Time (sec) | Number of Swings | Period (sec) |
|-------|------------|------------|------------------|--------------|
| 1 | | 60 | | |
| 2 | | 60 | | |
| 3 | | 60 | | |
| 4 | | 60 | | |

## Think for Yourself

You may have heard astronauts refer to the "gee forces" they experienced during lift-off. What do you think they were referring to?

# HOW CAN YOU OBSERVE THE SUN SAFELY?

As you know, it is extremely dangerous to look directly at the sun. Viewing the sun directly can result in permanent damage to your eyes. Is there a safe way to observe the sun? The answer is yes. The best way of looking at the sun is to project an image of the sun onto a piece of white paper. You can demonstrate this by making a simple pinhole viewer. You will need a shoe box, a white index card, tape, and a pin.

1. Tape the index card to the inside of one end of the shoe box. Use a pin to make a small hole in the opposite end of the shoe box. In a darkened room, hold the shoe box so that sunlight enters the pinhole. You should see an image of the sun projected onto the index card. Describe what you see.

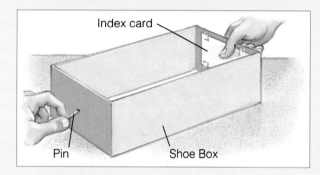

With a little simple mathematics, you can use a similar setup to measure the diameter of the sun. You will need a meterstick, two index cards, tape, and a pin.

2. Tape an index card to one end of the meterstick to make a screen. Make a pinhole in the other index card and hold it at the opposite end of the meterstick. Sunlight passing through the pinhole will form an image of the sun on the screen. Measure the diameter, in centimeters, of the sun's image on the screen. What is the diameter of the image?

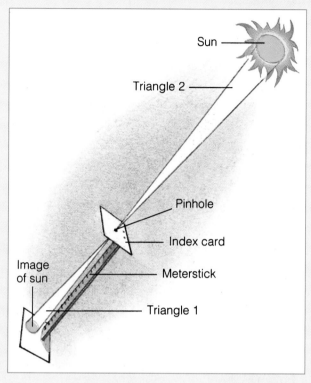

As you can see in the diagram, light rays passing through the pinhole to form the image make two similar triangles. This means that the ratio of the sun's diameter to its distance from the pinhole is the same as the ratio of the diameter of the image to the length of the meterstick. Use the following equation to calculate the sun's diameter:

Sun's diameter/150,000,000 km
= Image diameter/100 cm

What value did you find for the diameter of the sun?

## Think for Yourself

The Latin name for the pinhole viewer you made in this activity is *camera obscura,* which means dark chamber or room. Do you think this is an appropriate name for this device? Why or why not?

# RUSTY NAILS

Mars is often called the Red Planet. The surface of Mars appears red because the soil contains iron oxide—more commonly known as rust. You are probably familiar with rust closer to home. Anything made of iron that is exposed to air and moisture will become rusted. Junked cars, iron fences, and old bicycles are all subject to rusting. Is there any way to prevent objects from rusting? In this activity you will explore some ways to prevent rusting.

Water and vinegar

Coated with petroleum jelly

Coated with clear nail polish

Uncoated

## Materials

3 iron nails
clear nail polish
petroleum jelly
glass jar
vinegar

## Procedure

1. Coat one of the nails with clear nail polish. Coat the second nail with petroleum jelly. Do not put anything on the third nail.

2. Place the nails into a jar of water. Add some vinegar to the water to speed up the rusting process.

3. Allow the nails to stand in the glass jar overnight. Then examine the nails. Which nail shows signs of rusting? How do you think the nail polish and petroleum jelly prevented the nails from rusting?

## Going Further

What are some other substances that would prevent the nails from rusting? Repeat this experiment to test your ideas.

## Do It Yourself

Rusting can cause a great deal of damage to bridges and other objects made of iron by wearing away the metal. Rusty objects can also be dangerous to your health. If you accidentally cut yourself on a rusty nail or other sharp object, you should see a doctor immediately. Using first-aid books or other reference materials, find out why cuts caused by rusty objects are so dangerous.

# Activity Bank

## ACTION, REACTION

According to Newton's third law of motion, every action causes an equal and opposite reaction. This is the principle of reaction engines, such as rockets. It is also the principle that may cause you to get soaked if you try jumping from a small boat onto the dock! Here's a simple experiment you can perform to demonstrate Newton's third law of motion for yourself.

### Materials

skateboard
cardboard strip, 15 cm x 75 cm
windup toy car

### Procedure

1. Place the skateboard upside down on the floor.

2. Place the strip of cardboard on top of the wheels of the skateboard. The cardboard will be the "road" for the toy car.

3. Place the toy car on the cardboard, wind it up, and let it go. Observe what happens. Does the car or the road move?

### Think for Yourself

1. Are you aware of the road moving away from you when you are driving in a real car? Why or why not?

2. Would you be able to drive a car forward if you were not "attached" to the Earth?

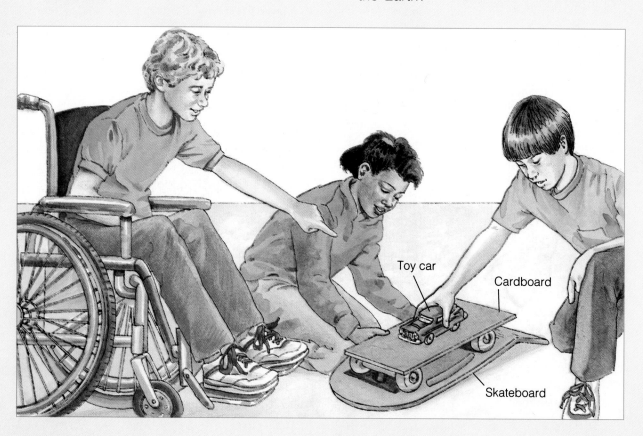

Toy car

Cardboard

Skateboard

# WHAT CAUSES HIGH TIDES?

The rise and fall of Earth's oceans—the tides—are caused by the pull of the moon's gravity on the Earth. Because the moon exerts different gravitational forces on different parts of the Earth, there are two high tides and two low tides every day at any given place. You can demonstrate the forces that cause the tides in this activity.

## Materials

construction paper
tape
drawing compass
3 equal masses
3 springs

## Procedure

1. Tape a piece of construction paper onto a smooth, flat surface. Using the compass, draw a circle 30 cm in diameter on the construction paper.

2. Label the three masses A, B, and C.

3. Attach the springs to the three masses as shown in the diagram. Place mass B in the center of the circle. The circle represents the Earth.

4. Apply a force to mass A by pulling on the spring. This force represents the gravitational pull of the moon on the Earth.

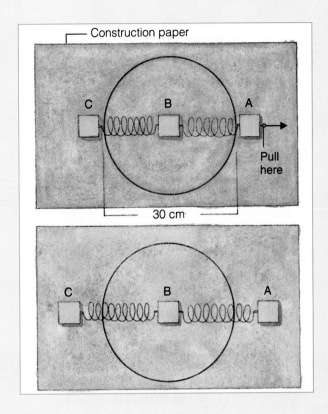

## Analysis and Conclusions

1. What happens to the other two masses when you exert a force on mass A?

2. How does this demonstration illustrate the two high tides on opposite sides of the Earth caused by the pull of the moon?

# Appendix A

# Appendix A

## THE METRIC SYSTEM

The metric system of measurement is used by scientists throughout the world. It is based on units of ten. Each unit is ten times larger or ten times smaller than the next unit. The most commonly used units of the metric system are given below. After you have finished reading about the metric system, try to put it to use. How tall are you in metrics? What is your mass? What is your normal body temperature in degrees Celsius?

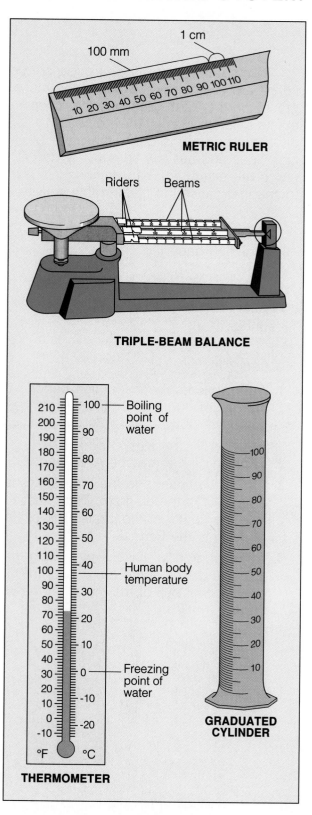

**METRIC RULER**

**TRIPLE-BEAM BALANCE**

**THERMOMETER**

**GRADUATED CYLINDER**

## Commonly Used Metric Units

### Length    The distance from one point to another

meter (m)    A meter is slightly longer than a yard.
1 meter = 1000 millimeters (mm)
1 meter = 100 centimeters (cm)
1000 meters = 1 kilometer (km)

### Volume    The amount of space an object takes up

liter (L)    A liter is slightly more than a quart.
1 liter = 1000 milliliters (mL)

### Mass    The amount of matter in an object

gram (g)    A gram has a mass equal to about one paper clip.

1000 grams = 1 kilogram (kg)

### Temperature    The measure of hotness or coldness

degrees     0°C = freezing point of water
Celsius (°C)    100°C = boiling point of water

## Metric–English Equivalents

2.54 centimeters (cm) = 1 inch (in.)
1 meter (m) = 39.37 inches (in.)
1 kilometer (km) = 0.62 miles (mi)
1 liter (L) = 1.06 quarts (qt)
250 milliliters (mL) = 1 cup (c)
1 kilogram (kg) = 2.2 pounds (lb)
28.3 grams (g) = 1 ounce (oz)
$°C = 5/9 \times (°F - 32)$

### Glassware Safety

1. Whenever you see this symbol, you will know that you are working with glassware that can easily be broken. Take particular care to handle such glassware safely. And never use broken or chipped glassware.
2. Never heat glassware that is not thoroughly dry. Never pick up any glassware unless you are sure it is not hot. If it is hot, use heat-resistant gloves.
3. Always clean glassware thoroughly before putting it away.

### Fire Safety

1. Whenever you see this symbol, you will know that you are working with fire. Never use any source of fire without wearing safety goggles.
2. Never heat anything—particularly chemicals—unless instructed to do so.
3. Never heat anything in a closed container.
4. Never reach across a flame.
5. Always use a clamp, tongs, or heat-resistant gloves to handle hot objects.
6. Always maintain a clean work area, particularly when using a flame.

### Heat Safety

Whenever you see this symbol, you will know that you should put on heat-resistant gloves to avoid burning your hands.

### Chemical Safety

1. Whenever you see this symbol, you will know that you are working with chemicals that could be hazardous.
2. Never smell any chemical directly from its container. Always use your hand to waft some of the odors from the top of the container toward your nose—and only when instructed to do so.
3. Never mix chemicals unless instructed to do so.
4. Never touch or taste any chemical unless instructed to do so.
5. Keep all lids closed when chemicals are not in use. Dispose of all chemicals as instructed by your teacher.

6. Immediately rinse with water any chemicals, particularly acids, that get on your skin and clothes. Then notify your teacher.

### Eye and Face Safety

1. Whenever you see this symbol, you will know that you are performing an experiment in which you must take precautions to protect your eyes and face by wearing safety goggles.
2. When you are heating a test tube or bottle, always point it away from you and others. Chemicals can splash or boil out of a heated test tube.

### Sharp Instrument Safety

1. Whenever you see this symbol, you will know that you are working with a sharp instrument.
2. Always use single-edged razors; double-edged razors are too dangerous.
3. Handle any sharp instrument with extreme care. Never cut any material toward you; always cut away from you.
4. Immediately notify your teacher if your skin is cut.

### Electrical Safety

1. Whenever you see this symbol, you will know that you are using electricity in the laboratory.
2. Never use long extension cords to plug in any electrical device. Do not plug too many appliances into one socket or you may overload the socket and cause a fire.
3. Never touch an electrical appliance or outlet with wet hands.

### Animal Safety

1. Whenever you see this symbol, you will know that you are working with live animals.
2. Do not cause pain, discomfort, or injury to an animal.
3. Follow your teacher's directions when handling animals. Wash your hands thoroughly after handling animals or their cages.

One of the first things a scientist learns is that working in the laboratory can be an exciting experience. But the laboratory can also be quite dangerous if proper safety rules are not followed at all times. To prepare yourself for a safe year in the laboratory, read over the following safety rules. Then read them a second time. Make sure you understand each rule. If you do not, ask your teacher to explain any rules you are unsure of.

## Dress Code

**1.** Many materials in the laboratory can cause eye injury. To protect yourself from possible injury, wear safety goggles whenever you are working with chemicals, burners, or any substance that might get into your eyes. Never wear contact lenses in the laboratory.

**2.** Wear a laboratory apron or coat whenever you are working with chemicals or heated substances.

**3.** Tie back long hair to keep it away from any chemicals, burners and candles, or other laboratory equipment.

**4.** Remove or tie back any article of clothing or jewelry that can hang down and touch chemicals and flames.

## General Safety Rules

**5.** Read all directions for an experiment several times. Follow the directions exactly as they are written. If you are in doubt about any part of the experiment, ask your teacher for assistance.

**6.** Never perform activities that are not authorized by your teacher. Obtain permission before "experimenting" on your own.

**7.** Never handle any equipment unless you have specific permission.

**8.** Take extreme care not to spill any material in the laboratory. If a spill occurs, immediately ask

your teacher about the proper cleanup procedure. Never simply pour chemicals or other substances into the sink or trash container.

**9.** Never eat in the laboratory.

**10.** Wash your hands before and after each experiment.

## First Aid

**11.** Immediately report all accidents, no matter how minor, to your teacher.

**12.** Learn what to do in case of specific accidents, such as getting acid in your eyes or on your skin. (Rinse acids from your body with lots of water.)

**13.** Become aware of the location of the first-aid kit. But your teacher should administer any required first aid due to injury. Or your teacher may send you to the school nurse or call a physician.

**14.** Know where and how to report an accident or fire. Find out the location of the fire extinguisher, phone, and fire alarm. Keep a list of important phone numbers—such as the fire department and the school nurse—near the phone. Immediately report any fires to your teacher.

## Heating and Fire Safety

**15.** Again, never use a heat source, such as a candle or burner, without wearing safety goggles.

**16.** Never heat a chemical you are not instructed to heat. A chemical that is harmless when cool may be dangerous when heated.

**17.** Maintain a clean work area and keep all materials away from flames.

**18.** Never reach across a flame.

**19.** Make sure you know how to light a Bunsen burner. (Your teacher will demonstrate the proper procedure for lighting a burner.) If the flame leaps out of a burner toward you, immediately turn off the gas. Do not touch the burner. It may be hot. And never leave a lighted burner unattended!

**20.** When heating a test tube or bottle, always point it away from you and others. Chemicals can splash or boil out of a heated test tube.

**21.** Never heat a liquid in a closed container. The expanding gases produced may blow the container apart, injuring you or others.

**22.** Before picking up a container that has been heated, first hold the back of your hand near it. If you can feel the heat on the back of your hand, the container may be too hot to handle. Use a clamp or tongs when handling hot containers.

## Using Chemicals Safely

**23.** Never mix chemicals for the "fun of it." You might produce a dangerous, possibly explosive substance.

**24.** Never touch, taste, or smell a chemical unless you are instructed by your teacher to do so. Many chemicals are poisonous. If you are instructed to note the fumes in an experiment, gently wave your hand over the opening of a container and direct the fumes toward your nose. Do not inhale the fumes directly from the container.

**25.** Use only those chemicals needed in the activity. Keep all lids closed when a chemical is not being used. Notify your teacher whenever chemicals are spilled.

**26.** Dispose of all chemicals as instructed by your teacher. To avoid contamination, never return chemicals to their original containers.

**27.** Be extra careful when working with acids or bases. Pour such chemicals over the sink, not over your workbench.

**28.** When diluting an acid, pour the acid into water. Never pour water into an acid.

**29.** Immediately rinse with water any acids that get on your skin or clothing. Then notify your teacher of any acid spill.

## Using Glassware Safely

**30.** Never force glass tubing into a rubber stopper. A turning motion and lubricant will be helpful when inserting glass tubing into rubber stoppers or rubber tubing. Your teacher will demonstrate the proper way to insert glass tubing.

**31.** Never heat glassware that is not thoroughly dry. Use a wire screen to protect glassware from any flame.

**32.** Keep in mind that hot glassware will not appear hot. Never pick up glassware without first checking to see if it is hot. See #22.

**33.** If you are instructed to cut glass tubing, fire-polish the ends immediately to remove sharp edges.

**34.** Never use broken or chipped glassware. If glassware breaks, notify your teacher and dispose of the glassware in the proper trash container.

**35.** Never eat or drink from laboratory glassware. Thoroughly clean glassware before putting it away.

## Using Sharp Instruments

**36.** Handle scalpels or razor blades with extreme care. Never cut material toward you; cut away from you.

**37.** Immediately notify your teacher if you cut your skin when working in the laboratory.

## Animal Safety

**38.** No experiments that will cause pain, discomfort, or harm to mammals, birds, reptiles, fishes, and amphibians should be done in the classroom or at home.

**39.** Animals should be handled only if necessary. If an animal is excited or frightened, pregnant, feeding, or with its young, special handling is required.

**40.** Your teacher will instruct you as to how to handle each animal species that may be brought into the classroom.

**41.** Clean your hands thoroughly after handling animals or the cage containing animals.

## End-of-Experiment Rules

**42.** After an experiment has been completed, clean up your work area and return all equipment to its proper place.

**43.** Wash your hands after every experiment.

**44.** Turn off all burners before leaving the laboratory. Check that the gas line leading to the burner is off as well.

## Boundaries

National . . . . . . . . . . . . . . . . . .

State or territorial . . . . . . . . . . . . . . .

County or equivalent . . . . . . . . . . . . .

Civil township or equivalent . . . . . . . .

Incorporated city or equivalent . . . . . .

Park, reservation, or monument . . . . .

Small park . . . . . . . . . . . . . . . . . . . . .

## Roads and related features

Primary highway . . . . . . . . . . . . . .

Secondary highway . . . . . . . . . . . . .

Light-duty road . . . . . . . . . . . . . . .

Unimproved road . . . . . . . . . . . . . .

Trail . . . . . . . . . . . . . . . . . . . . . . . .

Dual highway . . . . . . . . . . . . . . . . .

Dual highway with median strip . . . . . .

Bridge . . . . . . . . . . . . . . . . . . . . . . .

Tunnel . . . . . . . . . . . . . . . . . . . . . . .

## Buildings and related features

Dwelling or place of employment: small;

large . . . . . . . . . . . . . . . . . . . . . . .

School; house of worship . . . . . . . . . .

Barn, warehouse, etc.: small; large . . . .

Airport . . . . . . . . . . . . . . . . . . . . . . .

Campground; picnic area . . . . . . . . . .

Cemetery: small; large . . . . . . . . . . . .

## Railroads and related features

Standard-gauge single track; station . . .

Standard-gauge multiple track . . . . . . .

## Contours

Intermediate . . . . . . . . . . . . . . . . . . .

Index . . . . . . . . . . . . . . . . . . . . . . . .

Supplementary . . . . . . . . . . . . . . . . .

Depression . . . . . . . . . . . . . . . . . . . .

Cut; fill . . . . . . . . . . . . . . . . . . . . . .

## Surface features

Levee . . . . . . . . . . . . . . . . . . . . . . .

Sand or mud areas, dunes, or shifting

sand . . . . . . . . . . . . . . . . . . . . . . .

Gravel beach or glacial moraine . . . . . .

## Vegetation

Woods . . . . . . . . . . . . . . . . . . . . . . .

Scrub . . . . . . . . . . . . . . . . . . . . . . . .

Orchard . . . . . . . . . . . . . . . . . . . . . .

Vineyard . . . . . . . . . . . . . . . . . . . . . .

## Marine shoreline

Approximate mean high water . . . . . . . .

Indefinite or unsurveyed . . . . . . . . . . . .

## Coastal features

Foreshore flat . . . . . . . . . . . . . . . . . .

Rock or coral reef . . . . . . . . . . . . . . . .

Rock, bare or awash . . . . . . . . . . . . . .

Breakwater, pier, jetty, or wharf . . . . . . .

Seawall . . . . . . . . . . . . . . . . . . . . . .

## Rivers, lakes, and canals

Perennial stream . . . . . . . . . . . . . . . .

Perennial river . . . . . . . . . . . . . . . . . .

Small falls; small rapids . . . . . . . . . . . .

Large falls; large rapids . . . . . . . . . . . .

Dry lake . . . . . . . . . . . . . . . . . . . . . .

Narrow wash . . . . . . . . . . . . . . . . . . .

Wide wash . . . . . . . . . . . . . . . . . . . .

Water well; spring or seep . . . . . . . . . . .

## Submerged areas and bogs

Marsh or swamp . . . . . . . . . . . . . . . .

Submerged marsh or swamp . . . . . . . . .

Wooded marsh or swamp . . . . . . . . . . .

Land subject to inundation . . . . . . . . . .

## Elevations

Spot and elevation . . . . . . . . . . . . . . .    X$_{212}$

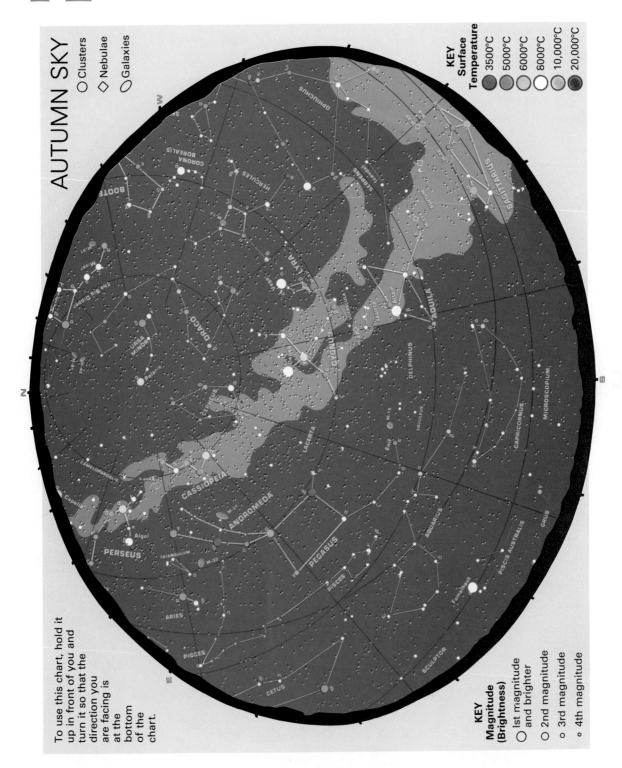

## AUTUMN SKY

○ Clusters
◇ Nebulae
○ Galaxies

To use this chart, hold it up in front of you and turn it so that the direction you are facing is at the bottom of the chart.

**KEY**
**Surface Temperature**
3500°C  5000°C  6000°C  8000°C  10,000°C  20,000°C

**KEY**
**Magnitude (Brightness)**
○ 1st magnitude and brighter
○ 2nd magnitude
○ 3rd magnitude
○ 4th magnitude

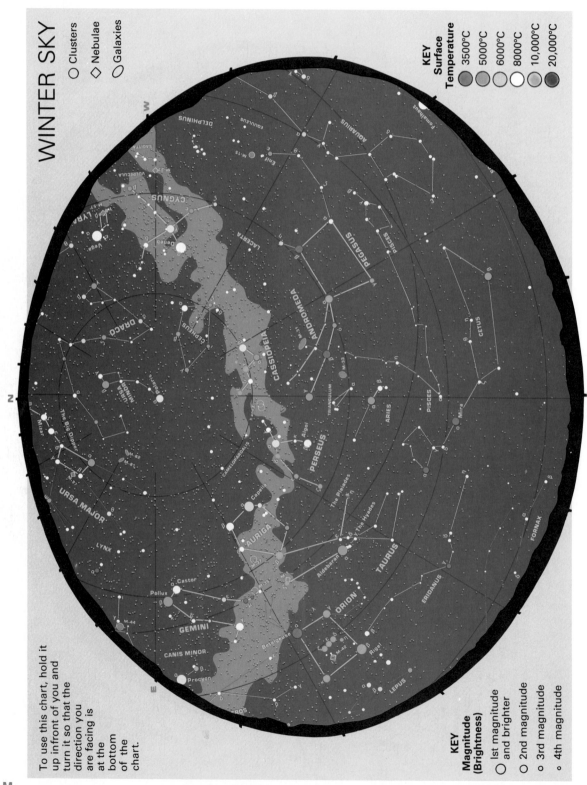

WINTER SKY

To use this chart, hold it up infront of you and turn it so that the direction you are facing is at the bottom of the chart.

○ Clusters
◇ Nebulae
⬭ Galaxies

**KEY**
**Surface**
**Temperature**
3500°C
5000°C
6000°C
8000°C
10,000°C
20,000°C

**KEY**
**Magnitude**
**(Brightness)**
○ 1st magnitude and brighter
○ 2nd magnitude
○ 3rd magnitude
∘ 4th magnitude

# Appendix E

STAR CHARTS

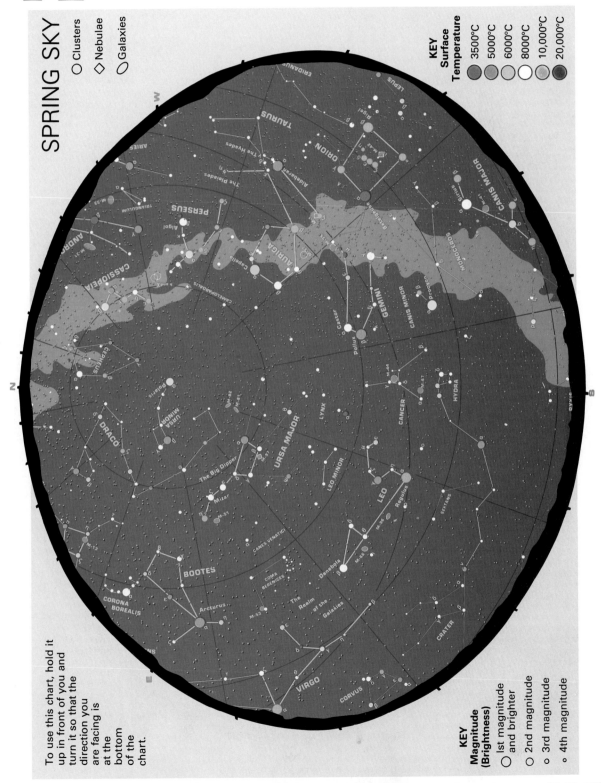

SPRING SKY

To use this chart, hold it up in front of you and turn it so that the direction you are facing is at the bottom of the chart.

Clusters ○
Nebulae ◇
Galaxies ⬭

KEY
Surface
Temperature
3500°C 5000°C 6000°C 8000°C 10,000°C 20,000°C

KEY
Magnitude
(Brightness)
◯ 1st magnitude and brighter
○ 2nd magnitude
∘ 3rd magnitude
· 4th magnitude

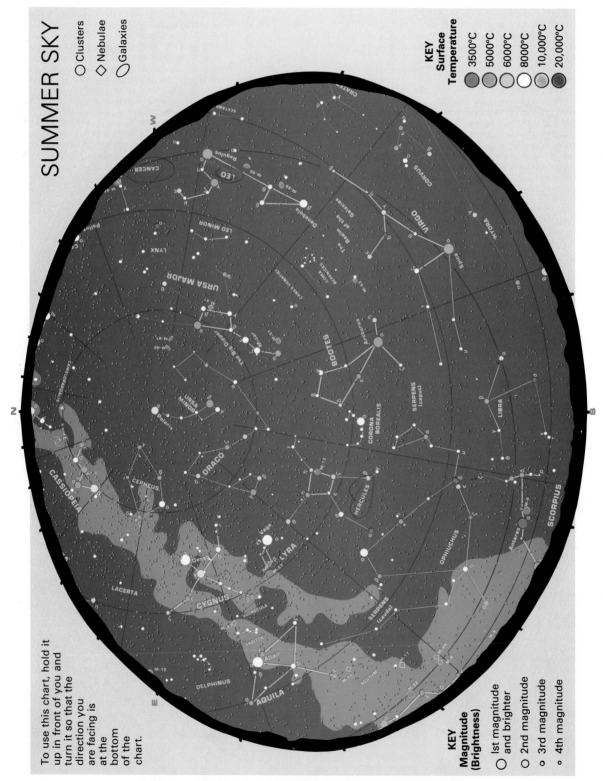

SUMMER SKY

Clusters ◇ Nebulae ◯ Galaxies

KEY
Surface
Temperature
3500°C 5000°C 6000°C 8000°C 10,000°C 20,000°C

To use this chart, hold it up in front of you and turn it so that the direction you are facing is at the bottom of the chart.

KEY
Magnitude (Brightness)
◯ 1st magnitude and brighter
◯ 2nd magnitude
◦ 3rd magnitude
· 4th magnitude

# Glossary

## Pronunciation Key

When difficult names or terms first appear in the text, they are respelled to aid pronunciation. A syllable in SMALL CAPITAL LETTERS receives the most stress. The key below lists the letters used for respelling. It includes examples of words using each sound and shows how the words would be respelled.

| Symbol | Example | Respelling |
|--------|---------|------------|
| a | hat | (hat) |
| ay | pay, late | (pay), (layt) |
| ah | star, hot | (stahr), (haht) |
| ai | air, dare | (air), (dair) |
| aw | law, all | (law), (awl) |
| eh | met | (meht) |
| ee | bee, eat | (bee), (eet) |
| er | learn, sir, fur | (lern), (ser), (fer) |
| ih | fit | (fiht) |
| igh | mile, sigh | (mighl), (sigh) |
| oh | no | (noh) |
| oi | soil, boy | (soil), (boi) |
| oo | root, tule | (root), (rool) |
| or | born, door | (born), (dor) |
| ow | plow, out | (plow), (owt) |

| Symbol | Example | Respelling |
|--------|---------|------------|
| u | put, book | (put), (buk) |
| uh | fun | (fuhn) |
| yoo | few, use | (fyoo), (yooz) |
| ch | chill, reach | (chihl), (reech) |
| g | go, dig | (goh), (dihg) |
| j | jet, gently, bridge | (jeht), (JEHNT-lee), (brihj) |
| k | kite, cup | (kight), (kuhp) |
| ks | mix | (mihks) |
| kw | quick | (kwihk) |
| ng | bring | (brihng) |
| s | say, cent | (say), (sehnt) |
| sh | she, crash | (shee), (krash) |
| th | three | (three) |
| y | yet, onion | (yeht), (UHN-yuhn) |
| z | zip, always | (zihp), (AWL-wayz) |
| zh | treasure | (TREH-zher) |

**absolute magnitude:** amount of light a star actually gives off

**apogee** (AP-uh-jee): point of a satellite's orbit farthest from the Earth

**apparent magnitude:** brightness of a star as it appears from Earth

**asteroid belt:** region of space between Mars and Jupiter in which asteroids are found

**aurora** (aw-RAW-ruh): bands or curtains of colored lights produced when particles trapped by the Van Allen radiation belts collide with particles in the upper atmosphere

**autumnal equinox** (EE-kwuh-naks): time of year when day and night are of equal length; beginning of autumn in the Northern Hemisphere

**axis:** imaginary vertical line through the center of body around which the body rotates, or spins

**big-bang theory:** theory that states that the universe began to expand with the explosion of concentrated matter and energy and has been expanding ever since

**binary star:** member of a double star system

**black hole:** core of a supermassive star that remains after a supernova; the gravity of the core is so strong that not even light can escape

**chromosphere** (KROH-muh-sfir): middle layer of the sun's atmosphere

**comet:** object made of ice, gas, and dust that travels through space

**constellation:** group of stars that form a pattern

**core:** center of the sun

**corona** (kuh-ROH-nuh): outermost layer of the sun's atmosphere

**Doppler effect:** apparent change in the wavelength of light that occurs when an object is moving toward or away from the Earth

**elliptical galaxy:** galaxy that may vary in shape from nearly spherical to flat; one of three types of galaxies

**escape velocity:** velocity needed to escape the Earth's gravitational pull

**galaxy:** huge collection of stars

**geosynchronous** (jee-oh-SIHNG-kruh-nuhs) **orbit:** orbit in which a satellite's rate of revolution exactly matches the Earth's rate of rotation

**giant star:** star with a diameter about 10 to 100 times as large as the sun

**gravity:** force of attraction between objects

**greenhouse effect:** process by which heat is trapped by a planet's atmosphere and cannot return to space

**Hertzsprung-Russell diagram:** chart that shows the relationship between the absolute magnitude and the surface temperature of stars; also called H-R diagram

**highlands:** mountain ranges on the moon

**lunar eclipse:** blocking of the moon that takes place when the Earth comes directly between the sun and the full moon

**magnetosphere:** magnetic field around a planet

**main-sequence star:** star that falls in an area from the upper left corner to the lower right corner of the H-R diagram

**maria** (MAHR-ee-uh; singular: mare): smooth lowland plains on the moon

**meteor:** streak of light produced by a meteoroid as it burns up in the Earth's atmosphere

**meteorite:** meteor that strikes the Earth's surface

**meteoroid** (MEE-tee-uh-roid): chunk of metal or stone that orbits the sun

**neap tide:** lower than usual high tide that occurs during the first and last quarter phases

**nebula:** massive cloud of dust and gas between the stars

**nebular theory:** theory which states that the solar system began as a huge cloud of dust and gas called a nebula, which later condensed to form the sun and its nine planets

**neutron star:** smallest of all stars

**nova:** star that suddenly increases in brightness in just a few hours or days

**nuclear fusion:** process by which hydrogen atoms are fused, or joined together, to form helium atoms

**orbit:** path an object takes when moving around another object in space

**parallax** (PAR-uh-laks): apparent change in the position of a star in the sky due to the change in the Earth's position as the Earth moves around the sun

**penumbra** (pih-NUHM-bruh): outer part of a shadow

**perigee** (PEHR-uh-jee): point of a satellite's orbit closest to the Earth

**period of revolution:** time it takes a planet to make one revolution around the sun

**period of rotation:** time it takes a planet to make one rotation on its axis

**photosphere:** innermost layer of the sun's atmosphere

**prominence** (PRAHM-uh-nuhns): violent storm on the sun that can be seen from Earth as a huge bright arch or loop of hot gas

**protostar:** new star

**pulsar:** neutron star that gives off pulses of radio waves

**quasar** (KWAY-zahr): quasi-stellar radio source; distant object that gives off mainly radio waves and X-rays

**reaction engine:** engine, such as a rocket, in which the rearward blast of exploding gases causes the rocket to shoot forward

**red shift:** shift toward the red end of the spectrum of a star that is moving away from the Earth

**retrograde rotation:** reverse motion in which a planet rotates from east to west, instead of from west to east

**rille:** valley on the moon

**solar eclipse:** blocking of the sun that occurs when the new moon comes directly between the sun and the Earth

**solar flare:** storm on the sun that shows up as a bright burst of light on the sun's surface

**solar system:** sun, planets, and all the other objects that revolve around the sun

**solar wind:** continuous stream of high-energy particles released into space in all directions from the sun's corona

**spectroscope:** instrument that breaks up the light from a distant star into its characteristic colors

**spectrum:** band of colors formed when light passes through a prism

**spiral galaxy:** galaxy that is shaped like a pinwheel; one of three types of galaxies

**spring tide:** higher than usual high tide that occurs during the full moon and new moon phases

**summer solstice** (SAHL-stihs): time of year when the Northern Hemisphere has its longest day and the Southern Hemisphere has its shortest day; beginning of summer in the Northern Hemisphere

**sunspot:** dark area on the sun's surface

**supergiant star:** star with a diameter up to 1000 times the diameter of the sun; largest of all stars

**supernova:** tremendous explosion in which a star breaks apart, releasing energy and newly formed elements

**tide:** rise and fall of the oceans caused by the moon's gravitational pull on the Earth

**umbra** (UHM-bruh): inner part of a shadow

**Van Allen radiation belts:** two doughnut-shaped regions of charged particles formed when the Earth's magnetosphere traps some of the particles in the solar wind

**vernal equinox** (EE-kwuh-naks): time of year when day and night are of equal length; beginning of spring in the Northern Hemisphere

**white dwarf:** small dense star

**winter solstice:** time of year when the Northern Hemisphere has its shortest day and the Southern Hemisphere has its longest day; beginning of winter in the Northern Hemisphere

# Index

Absolute magnitude, of stars, M33
Adams, John Couch, M83
Aerospace engineers, M119
Aldebaran, M29
Aldrin, Edwin, M113–114
Algol, M13–14
Alpha Centauri, M13
Andromeda, M18
Antares, M29
Apogee, moon, M117
*Apollo* missions, M113–114, M115
Apparent magnitude, of stars, M33
Ariel, M83
Armstrong, Neil, M113–114
Artificial satellites, M126–128
    communications satellites, M127
    navigation satellites, M127
    scientific satellites, M127–128
    weather satellites, M127
Asteroids, M59, M73–74
    asteroid belt, M59, M73–74
    characteristics of, M73
    collisions with Earth, M74
Astronomer's tools
    spectroscopes, M22
    telescopes, M22
Aurora, M111–112
    aurora australis, M112
    aurora borealis, M112
Autumn, M109
Autumnal equinox, M109
Axis
    Earth, M105
    and occurrence of seasons,
        M107–108
    sun, M42

Background radiation, M26
Barringer Meteorite Crater, M89
Beta Pictoris, M57
Betelgeuse, M15, M29
Big-bang theory, M25–26
    and closed universe, M26–27
    elements of, M25–26
    and gravity, M26
    and open universe, M26
    and quasars, M27
    second big-bang, M27
Big Bear, M15
Big Dipper, M15
Big Dog, M15
Binary stars, M13
Black holes, M48–49
    detection of, M48
    inside of, M49
Blue shift, M24, M27

Callisto, M78
Canis Major, M15
Canis Minor, M15
Carbon dioxide, and greenhouse
    effect, M69–70
Cepheid, M33
Cepheus, M33
Charon, M86
Christy, James, M86
Chromosphere, sun, M40
Closed universe concept, M26–27
Collins, Michael, M113
*Columbia*, M113
Comets, M59, M87–89
    Halley's comet, M88–89
    Oort cloud, M87–88
    parts of, M88
Communications satellites, M127
Constellations, M14–15
Copernicus, Nicolaus, M60–61
Copernicus crater, moon, M116
Core, sun, M40
Corona, sun, M40
Crab Nebula, M47, M48
Craters, moon, M116

Day, and rotation of Earth,
    M104–105
Dead stars, M46
Deep-space probes, M94–97
    *Magellan* exploration, M97
    *Mariner* explorations, M95
    *Pioneer* explorations, M94–95,
        M96
    *Voyager* explorations, M96
Deimos, M73
Distance to stars, M35–36
    parallax, M35–36
    spectroscope, M38
Dog Star, M14
Doppler effect, M24

*Eagle*, M113
Earth, M59
    axis, M105
    day and night, M104–105
    magnetic field, M110–112
    and moon, M113–125
    revolution of, M104, M106
    rotation of, M104–105, M117–118
    seasons, M107–109
    year, M106
Eclipses, M121–123
    lunar eclipse, M123
    penumbra, M122, M123

solar eclipse, M122–123
    umbra, M122, M123
Elliptical galaxies, M18–19
Elliptical orbits, M61–62
Equinox, autumnal and vernal,
    M109
Escape velocity, M93–94
Europa, M78
Evolution of solar system, M56–59
    formation of planets, M58–59
    formation of sun, M57–58
    nebular theory, M57
Evolution of stars
    black holes, M48–49
    massive stars, M46–47
    neutron stars, M47–48
    protostars, M43–44
    red giants, M44–45
    supernovas, M46–47
    white dwarfs, M46
*Explorer* explorations, M128

Faults, M83
*Freedom*, M129

Galaxies, M17–19
    elliptical galaxies, M18–19
    irregular galaxies, M19
    Milky Way, M19–20
    movement of, M23, M25
    spiral galaxies, M18
Galilei, Galileo, M77, M117
Galle, Johann, M83
Ganymede, M78
Gemini, M15
Geosynchronous orbit, M127
Giant stars, M29
Goddard, Robert H., M94
Gravity
    and big-bang theory, M26
    Newton's theory, M62
    and tides, M124–125
Great Red Spot, of Jupiter, M76,
    M80
Greenhouse effect, M69–70

Halley, Edmund, M88–89
Halley's comet, M88–89
Helium, and stars, M31, M44–45
Herschel, Sir William, M81
Hertzsprung, Ejnar, M33
Hertzsprung-Russell diagram,
    M33–35

Highlands, moon, M115–116
Hoba West meteorite, M89
*Hubble Space Telescope,* M57
Hunter, M15
Hydrogen, and stars, M31, M37, M43, M44–45

Inertia, law of, M62
Infrared Astronomy Satellite, M128
Infrared telescopes, M22
Io, M77–78
Irregular galaxies, M19

Jupiter, M59
    characteristics of, M75–79
    moons of, M77–78

Kant, Immanuel, M17–18
Kepler, Johannes, M61–62

Leap year, M106
Leo, M15
Leverrier, Jean Joseph, M82
Life, and solar system, M90–91
Life cycle of star. *See* Evolution of stars
Light, speed of, M12
Little Bear, M15
Little Dipper, M15
Little Dog, M15
Lunar eclipse, M123

*Magellan,* M55, M66, M95, M97
Magellan, Ferdinand, M97
Magnetic field, Earth, M110–112
Magnetosphere, M76–77, M110–111
    Van Allen radiation belts, M111
Maria, moon, M116
*Mariner* explorations, M64–65, M95
Mars, M59
    characteristics of, M71–74
    moons of, M73
Massive stars, M45–46
Mercury, M59, M62–63
    characteristics of, M64–66
Meteorites, M89–90
Meteoriods, M89–90
Meteors, M89
Milky Way, M11, M17, M18
    characteristics of, M19–20
*Mir,* M129
Miranda, M82
Mission to Planet Earth, M128
Moon, M113–125
    apogee, M117
    characteristics of, M114–115
    craters, M116

eclipses, M121–123
first men on, M113–114
highlands, M115–116
maria, M116
moon rocks, examination of, M115
movements of, M117–118
origin of, M118–119
perigee, M117
phases of, M120–121
rotation of, M117–118
and tides, M123–125
volcanic activity, M117
Motion, third law of, M92
Motions of planets
    Copernicus' theory, M60–61
    elliptical orbits, M61–62
    Kepler's theory, M61–62
    Newton's theory, M62
    period of revolution, M62–63
    period of rotation, M63
    Ptolemy's theory, M60
Multiple-star systems, M13–14

Navigation satellites, M127
Neap tides, M125
Nebulae, M16–17, M17, M43
Nebular theory, evolution of solar system, M57
Neptune, M59, M83
    characteristics of, M82–84
    moons of, M84
    rings of, M83
Neutron stars, M30, M47–48
Newton, Isaac, M62, M92
Night, and rotation of Earth, M104–106
Northern Hemisphere, M108, M109
North Pole, M105, M109, M110, M112
North Star, M14–15, M33
Novas, M15
Nuclear fusion
    and stars, M37, M44
    and sun, M37

*Olympus Mons,* M72
Oort, Jan, M87–88
Oort cloud, M87–88
Open universe concept, M26
Optical telescopes, M22
Orbit, M60
    elliptical orbits, M61–62
Orion, M15

Parallax, M35–36
Penumbra, eclipses, M122, M123
Perigee, moon, M117
Phases of moon, M120–122

Phobos, M73
Photosphere, sun, M40
*Pioneer* explorations, M94, M94–95, M96
*Pioneer Venus Orbiter,* M66
Planetary nebula, M46
Planets
    formation of, M58–59
    Jupiter, M75–79
    Mars, M71–74
    Mercury, M64–66
    motions of, M60–63
    Neptune, M82–84
    period of revolution, M62–63
    period of rotation, M63
    Planet X, M87
    Pluto, M85–86
    protoplanets, M59
    rocky planets, M59
    Saturn, M79–80
    and seasons, M107
    Uranus, M81–82
    Venus, M66–70
Planet X, M87
Pleiades, M16
Pluto, M59, M63
    characteristics of, M85–86
    moon of, M86
Polaris, M14–15, M33
Prominence, M40–41
Protoplanets, M59
Protostars, M43–44
Proxima Centauri, M13
Ptolemy, M60
Pulsars, M47–48

Quasars, M27–28
    and big-bang theory, M27

Radio telescopes, M22
Red giants, M34–35, M44–45
Red shift, M23–25
Revolution
    of Earth, M104, M106
    planet's period of, M62–63
Rigel, M15, M29
Rockets, M92–94
    escape velocity, M93–94
    history of, M92–94
    reaction engine, M92–93
Rocky planets, M59
Rotation
    of Earth, M104–105, M117–118
    planet's period of, M63
    retrograde rotation, M67
Russell, Henry Norris, M33

Satellites. *See* Artificial satellites
Saturn, M59
    characteristics of, M79–80

moons of, M80
rings of, M79–80
Scorpio, M15
Seasons, M107–109
  autumn, M109
  autumnal equinox, M109
  and planets, M107
  spring, M109
  summer, M108–109
  summer solstice, M109
  and tilt of axes, M107–108
  vernal equinox, M109
  winter, M109
  winter solstice, M109
Shapley, Harlow, M11, M17
Sirius, M14, M29
Skylab, M128–129
Solar eclipse, M122–123
Solar flares, M41
Solar storms, M113
Solar system
  comets, M87–89
  evolution of, M56–59
  life in, M90–91
  meteorites, M89–90
  meteoroids, M89–90
  meteors, M89
  motions of planets, M60–63
  planets, M58–87
Solar system exploration
  deep-space probes, M94–97
  rockets, M92–94
Solar wind, M41
Southern Hemisphere, M108,
  M109
South Pole, M105, M109, M110,
  M112
Space age
  artificial satellites, M126–128
  first exploration of space, M126
  space laboratories, M128–129
  space technology spinoffs,
    M130–131
Spacelab, M129
Space laboratories, M128–129
  *Freedom,* M129
  *Mir,* M129
  Skylab, M128–129
  Spacelab, M129
Space Shuttle, M129
Space technology spinoffs,
  M130–131
Spectroscope, M22–23, M23–24
  and composition of stars, M30–31
  distance to stars, M38
Spectrum, M23
Speed of light, M12
Spiral galaxies, M18

Spring, M109
Spring tides, M124–125
*Sputnik 1,* M126
Star clusters, M16
Stars
  absolute magnitude of, M33
  apparent magnitude of, M33
  binary stars, M13
  black holes, M48–49
  brightness of, M32–33
  color of, M32–33
  composition of, M30–31
  constellations, M14–15
  evolution (life cycle) of, M42–49
  galaxies, M17–19
  giant stars, M29
  Hertzsprung-Russell diagram,
    M33–35
  main-sequence stars, M34
  massive stars, M45–46
  measuring distance to, M35–36
  medium-sized stars, M44–45
  multiple-star systems, M13–14
  nebulae, M16–17
  neutron stars, M30, M47–48
  novas, M15
  and nuclear fusion, M37
  protostars, M43–44
  pulsars, M47–48
  red giants, M44–45
  shining of, M36–37
  size of, M28–30
  star clusters, M16
  sun, M38–42
  supergiant stars, M29
  supernovas, M46–47
  surface temperatures of, M31–32
  white dwarfs, M30, M45
Summer, M108–109
Summer solstice, M109
Sun, M38–42
  chromosphere, M40
  core, M40
  corona, M40
  formation of, M57–58
  layers of, M38–40
  and nuclear fusion, M37
  photosphere, M40
  prominence, M40–41
  rotation on axis, M42
  solar flares, M41
  solar wind, M41
  sunspots, M41–42
Sunspots, M41–42
Supergiant stars, M29, M35
Supernovas, M46–47
  events of explosion of, M46–47
  most famous, M47

Telescopes
  infrared telescopes, M22
  optical telescopes, M22
  radio telescopes, M22
  ultraviolet telescopes, M22
Tides, M123–125
  neap tides, M125
  spring tides, M124–125
Titan, M80
Tombaugh, Clyde, M85
Triton, M84
Tsiolkovsky, Konstantin E., M93

Ultraviolet telescopes, M22
Umbra, eclipses, M122, M123
Universe
  big-bang theory, M25–26
  closed universe concept, M26–27
  and movement, M23
  open universe concept, M26
  quasars, M27–28
  red shift, M23–25
Upper Atmosphere Research
  Satellite, M128
Uranus, M59
  characteristics of, M81–82
  moons of, M82
Ursa Major, M15
Ursa Minor, M15


*Valles Marineris,* M73
Van Allen, James, M111
Van Allen radiation belts, M111
Van Maanen's star, M30
*Venera* explorations, M66
Venus, M55, M59
  characteristics of, M66–70
Vernal equinox, M109
*Viking* explorations, M71–72, M90,
  M95
Virgo, M15
*Voyager* explorations, M77, M80,
  M82, M83, M96
Weather satellites, M127
White dwarfs, M30, M45
Winter, M109
Winter solstice, M109


Year
  Earth, M106
  leap year, M106


Zodiac, M21

Credits

**Cover Background:** Ken Karp
**Photo Research:** Natalie Goldstein
**Contributing Artists:** Illustrations: Warren Budd Assoc. Ltd.; Kim Mulkey; Mark Schuller. Charts and graphs: Function Thru Form
**Photographs: 4** left: Mike James/Photo Researchers, Inc.; right: US Naval Observatory/Science Photo Library/Photo Researchers, Inc.; **5** top: NASA; bottom: Tom Bean/DRK Photo; **6** top: Lefever/Grushow/ Grant Heilman Photography; center: Index Stock Photography, Inc.; bottom: Rex Joseph; **8** top: Bill Iburg/Milon/Science Photo Library/Photo Researchers, Inc.; bottom: NASA; **9** NASA/JPL Photo/The Planetarium, Armagh, N. Ireland; **10** and **11** US Naval Observatory/Science Source/Photo Researchers, Inc.; **12** George East/Science Photo Library/Photo Researchers, Inc.; **13** R. B. Minton; **15** Lick Observatory; **16** left: NASA/US Naval Observatory; right: Hale Observatories; **17** left: D. Malin, Anglo-Australian Telescope Board/The Plane-tarium, Armagh, N. Ireland; right: Ralph Mercer/Tony Stone Worldwide/ Chicago Ltd.; **18** top: Dr. Jean Lorre/Science Photo Library/Photo Researchers, Inc.; center: Science VU/Visuals Unlimited; bottom left: US Naval Observatory/Science Photo Library/Photo Researchers, Inc.; bottom right: Science VU-NO/Visuals Unlimited; **19** top: D. Malin, Anglo-Australian Telescope Board/The Planetarium, Armagh, N. Ireland; bottom: Dr. William C. Keel/Science Photo Library/Photo Researchers, Inc.; **20** Lund Observatory; **22** top left: Fred Espenak/Science Photo Library/Photo Researchers, Inc.; right: NASA/Science Source/Photo Researchers, Inc.; bottom left: Max-Planck-Institut Fur Physik und Astrophysik/Science Photo Library/Photo Researchers, Inc.; **23** Phototake; **27** Center for Astrophysics; **30** left: E.Helin/JPL/NASA; right: Fred Espenak/Science Photo Library/Photo Researchers, Inc.; **32** Ronald Royer/Science Photo Library/Photo Researchers, Inc.; **36** left: Max-Planck-Institute for Radio Astronomy/Science Photo Library/Photo Researchers, Inc.; right: NOAO/Science Photo Library/Photo Researchers, Inc.; **39** AP/Wide World Photos; **40** Phototake; **41** NASA/JPL Photo/The Planetarium, Armagh, N. Ireland; **42** Science Photo Library/Photo Researchers, Inc.; **43** Tom Bean/DRK Photo; **44** Stock Market; **45** Lick Observatory; **46** top and center: D. Malin, Anglo-Australian Telescope Board/The Planetarium, Armagh, N. Ireland; bottom: Goddard Space Flight Center/NASA; **47** Science VU/Visuals Unlimited; **49** NASA; **54** and **55** Jet Propulsion Laboratory; **56** NASA; **61** Hansen Planetarium; **64** NASA; **66** NASA/LBJ Space Center; **67** NASA/JPL Photo/The Planetarium Armagh, N. Ireland; **70** left: Jet Propulsion Laboratory; top right and bottom right: NASA; **71** top: NASA; bottom: Jet Propulsion Laboratory; **72** top: JPL/CIT/NASA; bottom: Jet Propulsion Laboratory; **73** NASA/JPL Photo/The Planetarium, Armagh, N. Ireland; **76, 77, 78,** and **79** NASA; **80** top and center: NASA; bottom: Jet Propulsion Laboratory; **81** and **82** Jet Propulsion Laboratory; **83** NASA/JPL/The Planetarium, Armagh, N. Ireland; **84** NASA; **86** top: NASA/ JPL/The Planetarium, Armagh, N. Ireland; bottom left and right: NASA and ESA/The Planetarium, Armagh, N. Ireland; **89** top left: © 1978 AURA, Inc., Kitt Peak National Observatory; top right: British Museum/ National Air & Space Museum, Smithsonian Institution; bottom: NASA; **90** John Sanford/Science Photo Library/Photo Researchers, Inc.; **92, 94, 95, 101, 102, 103, 104,** and **105** NASA; **109** top left: Francois Gohier/Photo Researchers, Inc.; top right: Mike James/Photo Researchers, Inc.; bottom: Jon Riley/Folio, Inc.; **110** William E. Ferguson; **112** NASA; **113** Jose Azel/ Woodfin Camp & Associates; **114, 115, 116, 117,** and **118** NASA; **119** M. Polak/Sygma; **121** Lick Observatory; **122** Astro-Physics, Inc.; **125** Everett C. Johnson/Folio, Inc.; **126** NASA; **127** top: NASA; bottom: NOAA; **128** left: E.R.I.M./Tony Stone Worldwide/ Chicago Ltd.; right: Sygma; **129** top: NASA; bottom right: Matson Press and Mashnostronie Press/© 1991 Discover Magazine; **130** NASA; **136** Vera Lentz/ Black Star; **137** left: AP/Wide World Photos; right: NASA; **138** Movie Still Archives; **139** Roger Ressmeyer/Starlight; **140** Seth Resnick; **144** top and bottom, and **161** NASA; **164** Jon Riley/Folio, Inc.